Environments

Full Option Science System
Developed at
The Lawrence Hall of Science,
University of California, Berkeley
Published and distributed by
Delta Education,
a member of the School Specialty Family

1487707
978-1-62571-336-0
Printing 2 — 6/2016
Quad/Graphics, Leominster, MA

Table of Contents

Two Terrestrial Environments

Environmental scientists know a lot about Earth's **environments**. There are **aquatic** environments and **terrestrial** environments. *Terrestrial* refers to Earth's land. There are six major terrestrial environments in the world. They are tropical rain forest, desert, temperate deciduous forest, grassland, taiga, and tundra.

Each environment can be described in terms of **environmental factors**. Environmental factors are **living** and **nonliving** parts of the environment. The living parts of an environment are all the plants and animals that live and **thrive** in that place. The main nonliving components that define the six environments are **temperature**, rainfall, and soil type.

The tropical rain forest environment is different from the desert environment. The tropical rain forest is hot and wet, and the soil is poor because it lacks **nutrients**. The desert environment is dry and sandy. Most deserts are hot, but some are cold. Let's take a closer look at these two terrestrial environments and compare the living and nonliving environmental factors.

A tropical rain forest environment and a desert environment

Living Factors in Tropical Rain Forests

Tropical rain forests are home to more kinds of life than any other terrestrial environment. At least half of all the different kinds of plants and animals in the world live in tropical rain forests. Tropical rain forests are also the winter homes for many birds that live in other places the rest of the year.

Life in the rain forest can be divided into layers. Each layer has different plants and animals. Most of the tropical rain forest plants are trees. They grow to heights of 20 to 30 meters (m). Because the trees grow very close to one another, their tops grow together. This forms a broad **canopy**, or roof, above the rain forest.

A tropical rain forest canopy

The highest layer in the rain forest is the canopy. There is a lot of sunlight in the canopy layer. This is where most of the rain forest animals live. Monkeys, sloths, and bats spend most of their time here. Tree frogs and snakes live in the treetops along with toucans, hummingbirds, ants, and beetles. These are just a few of the millions of different kinds of animals that live in the canopy. Orchids, ferns, and other "air plants" grow on the branches of the canopy trees. Air plants use the trees for support and get water from the falling rain.

The layer below the canopy is the **understory**. Very little sunlight makes it through the canopy to the understory. It is a dark place full of tree trunks, young thin trees, and broad-leafed plants that thrive in shady conditions. A number of these plants are popular house plants in the United States. The animals living in this layer include jaguars, leopards, frogs, snakes, parakeets, and many kinds of **insects**.

The bottom layer is the forest floor. The forest floor is often covered with moss and wet leaves. Almost no sunlight makes it to the floor. This is where centipedes and scorpions live. Many insects, such as termites, ants, cockroaches, and beetles, also live here. Earthworms and **fungi** use the dead leaves as **food**. Larger animals, such as tapirs, dig up roots in the forest floor.

Can you identify the tapir, frog, and sloth?

Living Factors in Deserts

Some people think of deserts as hot, dry wastelands. That can be true, sometimes. In areas of shifting sand where it never rains and strong winds blow, such as parts of the Sahara Desert in Africa, plants and animals are rare. But deserts have areas that get some water, and those areas are full of life.

Fewer kinds of plants and animals live in deserts than in wetter environments. Desert plants and animals have **structures** and **behaviors** that help them survive in a dry environment. You can see plants and animals with these adaptations in parts of the deserts found in the southwestern United States.

Sand dunes in the Sahara Desert in northern Africa

The Mojave Desert in the American southwest

In deserts, some plants grow far apart. Their root systems spread over a large area. This distance lets them get water and nutrients without competition from other plants. Some desert plants, such as the mesquite tree, send their roots deep into the desert soil. Mesquite tree roots might go down 81 m to reach water.

Cacti store water in their broad, fleshy blades or columns, which are actually stems. They use the stored water during long dry periods. Cacti don't have leaves but they do have spines. The seeds of some desert plants can lie in the soil for years until it rains enough for them to sprout.

A saguaro cactus

A desert bighorn sheep

A desert tortoise

A desert iguana

Animals survive well in the southwest deserts. Insects, spiders, reptiles, birds, and mammals, such as bighorn sheep, live in deserts. Many desert animals are **nocturnal**. Nocturnal animals avoid the heat by coming out only at night.

Desert tortoises are comfortable in the desert. They dig deep **burrows**. When it is too hot or too cold, they have a safe place to stay. Tortoises eat many kinds of plants, especially flowers and fruits. Sometimes they will even eat the moist pads of cactus plants. Tortoises drink a lot of water when they can and store it in their bladders.

A spadefoot toad

Spadefoot toads are **amphibians**. That means they have to **reproduce** in water. Is the desert a good place for them to live? Yes, because they have a behavior to help them survive. When the weather is hot and dry, the toads burrow about a meter underground. They can stay there for up to 9 months. They become **dormant** and live on the fat stored in their bodies. When it rains, spadefoot toads leave their burrows and find mates. The females lay eggs in rain puddles. The eggs soon hatch into tadpoles. The tadpoles grow into young toads. The young toads have to become adults before the puddles dry up or they will die. In a couple of months, the adult toads burrow down into the ground and wait for next year's rain.

Every desert plant and animal has structures and behaviors that allow it to survive and thrive in the hot, dry desert.

The stem of a barrel cactus is round with ribs that are covered with spines.

9

Nonliving Factors in Rain Forests

Look at the map to see where tropical rain forests are on Earth. Can you find the ones in Australia? In Asia? In Africa? In Central America? In South America? Where else are there tropical rain forests? Find the line that shows the equator.

Tropical rain forests are found near the equator.

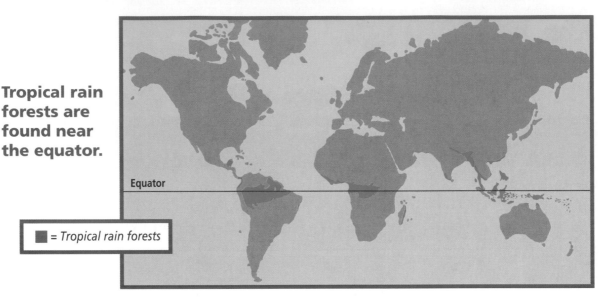

Equator

■ = *Tropical rain forests*

Tropical rain forests are wet and hot all year. The rainfall in rain forests is about 200 to 450 centimeters (cm) per year. How does that compare to where you live? Here are average rainfalls for five cities in the United States.

- Houston, Texas = 122 cm
- Charlotte, North Carolina = 110 cm
- Chicago, Illinois = 92 cm
- Anchorage, Alaska = 40.5 cm
- Phoenix, Arizona = 21.5 cm

The rain forest soil is shallow and not very **fertile**. Most of the nutrients that plants need to survive are in the trees. If the trees are cut down and taken away, the nutrients are lost to the rain forest environment. This is why it takes a long time for tropical rain forests to grow back once they are destroyed.

El Yunque National Forest

Nonliving Factors in Deserts

Scientists define a desert environment as any place on Earth that receives less than 25 cm of rain per year. Soils are rocky or sandy in deserts. Water runs off the land quickly or sinks into the sand. Water **evaporates**, or dries up, quickly in the desert. Most of the small amount of water that does fall on the desert is lost before plants and animals can get to it. Look at the map to see where deserts are on Earth.

Deserts are found north and south of the equator.

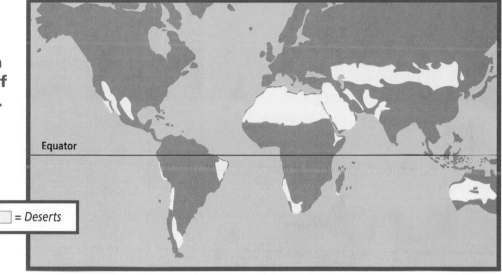

Equator

☐ = Deserts

Deserts are the hottest places on Earth during the summer. But during the winter, the temperatures can drop below freezing. Snow is seen regularly in parts of the deserts in southern California, China, and South America.

About 20 percent of Earth's land surface is desert. The small amount of rain, high temperatures, and large temperature changes from season to season make life challenging in the desert.

Snow on Joshua trees in the Mojave Desert

11

A tropical rain forest environment   **A desert environment**

Thinking about Environmental Factors

1. What are the environmental factors that define a tropical rain forest environment?

2. What are the environmental factors that define a desert environment?

3. What are some of the structures and behaviors that help organisms survive in the desert?

Setting Up a Terrarium

People enjoy having plants around. Some people grow plants in gardens. Others grow plants in pots in their homes. Some people grow plants in **terrariums**. A terrarium is a container with plants growing inside. A terrarium can also be a **habitat** for small animals found in a garden.

Terrariums can be any size and shape. Any clear container can be made into a terrarium. It should have a deep layer of soil and a lid to keep moisture inside. It also needs to be large enough to hold the plants and animals you want to keep inside it. A well-planned terrarium provides a good environment for living things.

So what makes a good terrarium environment? All **organisms** have needs. Plants need air, water, nutrients, light, space, and the proper temperature. Animals need air, water, food, space, shelter, and the proper temperature. Plants can be a very good source of food or shelter for some animals.

Not every kind of plant needs the same environment. Some plants need a lot of water. Others require only a small amount. Some plants need bright light, while others thrive in shade. Some plants grow best in cool temperatures, and others thrive in heat. Plan your terrarium to suit the organisms that will live in it.

Terrariums are many sizes and shapes.

Plants need light and space to grow.

An organism's environment is *everything* that surrounds and affects it. Each part or component of an organism's environment is an environmental factor. An environmental factor can be nonliving, such as water, light, and air temperature. Environmental factors also can be living, like all the plants and animals surrounding an organism.

So, when making a terrarium, you need to think about the environmental factors for the plants and animals that will live in it. What kind of soil will you use? Will you have one kind of plant or several kinds? How much water will you provide for the organisms and how often? How will air flow through the container? Where should the terrarium be placed in the room for light and temperature?

A terrarium can be created to represent a desert, grassland, woodland, or rain forest environment. You can add small animals that live in these natural habitats to the terrarium and observe them over time. Reptiles, amphibians, insects, **crustaceans**, worms, and spiders are good animals to place in a terrarium.

Water is a nonliving environmental factor.

Homemade Terrarium

You can make your own terrarium at home using a **recycled** 2-liter (L) clear plastic bottle. Here's how.

What You Need

1 2 L plastic bottle (colorless, clear)
- Soil
- Gravel or small pebbles
- Seeds or small plants
- Scissors
- Water
- Small garden animals
- Piece of carrot

What You Do

1. Remove the label from the plastic bottle. With help from an adult, cut the bottle about 10 centimeters (cm) from the bottom. Leave the cap on the bottle.

2. Cut four 2 cm slits along the bottom edge of the top part of the bottle.

3. Put a layer of gravel or small pebbles in the base. Add a layer of soil. If you are planting seeds in the terrarium, fill the base with soil almost to the top. Then plant your seeds. If you are planting a small rooted plant, dig a hole in the soil. Place the roots of the plant in the hole. Fill soil in around the roots. Water the soil.

4. Add small garden animals, such as earthworms or isopods. Put a piece of carrot in the terrarium for food.

5. Place the top section of your bottle on the base, fitting the slits over the base. Place the terrarium in an area with some light. Observe how your organisms grow in their environment.

Thinking about Words

The word *terra* means "earth" or "land." The suffix *-arium* means a place. What do you think *aquarium* means?

Isopods

Pill bugs! Sow bugs! These are two common names people give to **isopods**. Some people confuse isopods with insects. Isopods are not insects because isopods have seven pairs of legs. All insects have only *three* pairs of legs.

All seven pairs of an isopod's legs have the same **function**. Their scientific name indicates this: *iso* means "similar" or "equal," and *pod* means "foot." Isopods use all seven pairs of legs for walking, and nothing else. Insect legs are used for many functions. These include feeding, grasping, jumping, swimming, and carrying. This is another way that isopods are different from insects.

Isopods

Pill bugs

There are many kinds of isopods, but all are crustaceans. Crustaceans are animals with shells, jaws, and two pairs of antennae. Crustaceans include crabs, shrimp, and lobsters. Most crustaceans live in water and breathe with gills. Isopods are a little different. They can live on land. But they have to be in a moist environment most of the time. As long as they keep their gill-like breathing structures wet, they can breathe. If these structures become dry, the isopod cannot survive. The gills are located behind the last pair of walking legs on the isopod's underside.

Did you observe two different kinds of isopods in class? One kind is dome-shaped and has short antennae. When this isopod senses danger, it can roll up into a ball. That's why it is called a pill bug.

The other isopod is flatter and has longer antennae. It is called a sow bug. A sow bug cannot roll up to protect itself from a hungry spider or insect. But it can run faster than a pill bug.

Sow bugs and pill bugs feed on dead leaves and decaying fruit and seeds. They play an important role in recycling dead plant material in many environments. Where have you found isopods?

A sow bug

Amazon Rain Forest Journal

My name is Lee. My mother is an **entomologist**. That's a biologist who studies insects. She has traveled to the Amazon River in Brazil many times. Her stories about the rain forest always sounded so exciting. I begged Mom for a long time to take me on one of her trips to the rain forest. Finally, she surprised me with a special birthday trip. I wrote a journal about my trip to share my experiences with everyone at school.

Monday, July 21

It's early morning, and I am sitting in the Manaus airport in Brazil. I can't believe that I'm so far from home in Austin, Texas. Home is 5,472 kilometers (km) away!

We are waiting for a small plane to take us down the Amazon River to the city of Santarém. That's where the ecology research station is. Mom is one of more than 100 scientists who works there.

It's early afternoon now. The flight was great. The rain forest below looked like a green carpet as far as I could see. Mom's friend Kopenawa met us at the airport. She has known Kopenawa for a long time. He has guided her safely through the rain forest many times. He is ready to take us for a short hike in the rain forest. Mom says he knows more about the rain forest environment than almost anyone else.

Manaus

A green lizard can blend in with plants.

We're going down the biggest river in the world. Once the boat moves away from the city, I know I am in a new environment. The air is really hot and humid. I can feel sweat soaking my clothes. Everything I see is huge! The trees are the tallest I've ever seen, and there are interesting plants with gigantic leaves everywhere I look.

After traveling a short distance, Kopenawa guided the boat over to the riverbank. We're going into the rain forest!

We are back in Kopenawa's boat after a short hike in the rain forest. From the riverbank, the rain forest looked really dense. I didn't think we would be able to walk through it. But it was much easier to walk when we got away from the river and into the trees. Mom said this is because the tops of the trees form a canopy, or cover. Only a little light can make it through the canopy, so few plants actually grow on the forest floor.

I saw my first rain forest animal. It was a big green lizard! I nearly stepped on it because it blended in so well with the plants. Kopenawa said the lizard's camouflage makes it very hard for **predators** to see. It was hard for me to see, too.

Leaf-cutter ants

Tuesday, July 22

After our short hike in the rain forest yesterday, Kopenawa took us to the research station. We have a little room in a cabin with just screens on the windows. The sounds of the birds and monkeys woke me up early. I ate breakfast in a room full of long tables. Scientists were talking about rainfall, soil, temperature, seed sprouting, parrots, beetles, and a lot more. I was ready to find out more for myself.

Near the cabin I saw a line of thousands of very large ants. They were marching through the rain forest on a trail they made. When I got closer, I could see that many of the ants held pieces of green leaves over their heads. They looked like they were carrying tiny umbrellas. They really looked funny. I asked Kopenawa if they were protecting themselves from the rain. He laughed and told me that they were leaf-cutter ants. They cut pieces out of leaves and carry them back to their underground nests. Their nests are made of hundreds of small rooms, called chambers, under the earth.

The leaf-cutter ants don't eat the leaves. In fact, some of the leaves are poisonous. The ants use the leaves to grow a type of fungus. They chew the leaves. This makes a bed of leaf pulp where the fungus grows. The fungus is what the ants eat.

I asked Kopenawa if the ants ever get lost in the rain forest. The ground is covered with roots, rocks, and plants. I thought they must have trouble finding their way home. Kopenawa explained that the ants put down drops of a chemical, called a pheromone. The pheromones mark the trail for other ants to follow.

These ants use leaves to grow their own food.

Red army ants

Leaf-cutter ants reproduce with **complete metamorphosis**, like all ants. The queen ant lays eggs that hatch into **larvae**. The adult ants feed fungus to the larvae until the larvae pupate. Soon after, the adults come out. The new ants are called workers. They get right to work cutting leaves and growing fungus.

We kept walking. In half an hour, Kopenawa stopped and pointed to a spot near the trail. More ants! But this time, a battle was going on. A group of red army ants was attacking a group of wasps. Army ants do not eat leaves or fungus. They eat other insects such as wasps, moths, and grasshoppers. Wasps and ants are called social insects because they live together in groups. We watched the ants swarm over the wasps' nest. There were so many army ants that the wasps could not defend themselves. The wasps could only fly away and leave their larvae and eggs behind. The ants carried the wasp eggs and larvae back to their own nest. The eggs and larvae would be food for the rest of the army ants.

A short distance away, we saw a ball of army ants the size of a basketball. Kopenawa said the ants were making a temporary nest. This is the only kind of nest they ever make. The ants hook themselves together in chains. The chains form chambers for the queen and growing larvae. It was like something out of a science fiction movie, a living fort.

Mom told me that the army ants have to keep moving. They eat all the insects and other small animals in their path. They need to keep finding new places to get food.

A coral snake

Wednesday, July 23

Today we got up very early and went for a hike right after breakfast. Suddenly, Kopenawa put out his hand and stopped me in my tracks. A coral snake was slithering through the leaves on the ground. The snake had bright bands of yellow, red, and black. Some rain forest animals are brightly colored. Often the brightly colored animals are poisonous. The bright colors warn predators to stay away. I kept my distance. Then the snake disappeared into the rain forest.

Thursday, July 24

Today I went with my mom to her study area. She studies plants and the insects that eat them. I was surprised to hear that most of the trees we were walking under were poisonous. Mom said that's how trees defend themselves against the millions of hungry insects in the rain forest. If the plants didn't have defensive chemicals, all their leaves would be eaten and they would die.

Mom studies why some insects are not affected by the poisonous leaves. Every kind of tree seems to be eaten by one or two kinds of insects. Why can those insects eat leaves that are poisonous to every other kind of insect? That's what Mom tries to figure out.

Mom pointed out a small tree and warned me not to touch it. I thought it might be poisonous, but Mom said no, it was the ants that lived on it. Because this small tree grows in the rain forest understory, it doesn't get much light. It needs every leaf to survive. The ants attack any leaf-eating animals that come close to the tree. The ants make it possible for the tree to survive.

Ants live inside these large, hollow thorns.

The small tree has hollow thorns that provide shelter for the ants. The tree also feeds a number of sap-sucking insects called aphids. The aphid herds produce a sweet substance called honeydew. The ants feed on the honeydew. The tree makes it possible for the ants to survive.

Seeing the ant tree reminded Mom of a tree called the swollen-thorn acacia. She observed it on a study trip to the rain forest in Costa Rica in Central America. The leaves of this acacia tree are not poisonous, but the tree is not eaten by insects. It has another way to survive.

The acacia tree produces sugar syrup and little fruitlike bulbs. The bulbs are rich in vitamins and proteins. But only one kind of insect, an ant, eats the abundant food. Why?

Again, the ants protect the tree! When a hungry insect lands on the acacia tree, the ants attack it. If a vine touches the tree, the ants chew through the vine and cut it from the tree. As the acacia tree grows, the ants cut away the ends of the branches on the neighboring trees. The large, swollen acacia thorns are hollow. The ants live safely inside the thorns.

It's amazing. The acacia tree provides food and protection for the ants, and the ants protect the tree. The tree and the ants depend on each other for survival.

Ants sipping sugar syrup

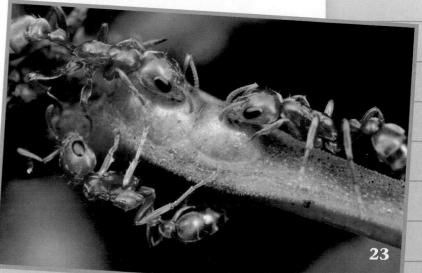

Friday, July 25

This was the best day of all. I had been looking forward to it all week. I actually went into the top of the rain forest today, up into the canopy. The research station has a set of walkways in the canopy. The trip to the canopy started with an elevator ride. Mom called it a "lift." The lift carried us up 30 meters (m). When the lift stopped, we stepped onto a maze of walkways that stretched through the treetops. Each tree trunk had a strong wooden platform around it. The platforms were big enough for a few people to stand on them. As I looked around, I felt like I was on top of the world. We were in the rain forest canopy.

It would be hard to describe all the animals I saw. I observed many kinds of birds. I saw colorful macaws and toucans. Many of the trees were covered with small berries. The branches were full of howler monkeys, dwarf squirrels, and tree frogs. I was amazed that the top of the rain forest was so full of life. It was very different from the forest floor.

A toco toucan

A red-eyed tree frog

A scarlet macaw

A howler monkey

24

Saturday, July 26

This was my last day in the rain forest. I spent so much time hiking around that I am too tired to write very much. Even though the weather was hot and humid and it rained every day of the trip, I didn't mind. There were so many new things to see and hear!

I'll miss Kopenawa. It was hard to say goodbye to him today. And I'll never forget all the animals, the toucans, howler monkeys, snakes, and the rest. It was amazing to see so many different kinds of plants and animals living together. Everything seems to fit together and work together. It's like fitting all the pieces of a puzzle together to make a beautiful picture. I'll keep this picture in my mind all the way back to Austin.

The rain forest canopy

The canopy walkway

25

Studying the Rain Forest

Lee's mother is an entomologist. She studies insects in the rain forest. These organisms, like many plants, animals, and fungi, can survive only in the rain forest. The cutting of trees is destroying the rain forest and causing the **extinction** of many plants and animals. Little is known about many of the organisms in the rain forest. Studying them is important because of the potential benefits this knowledge may bring to the world.

Thinking about Rain Forests

1. What did Lee learn about ants on the rain forest adventure?

2. How do ants communicate with each other about navigating through the rain forest?

3. In what ways do animals depend on plants in the rain forest environment? How do the plants depend on the animals in the rain forest environment?

4. What environmental factor changes as you go from the rain forest canopy to the rain forest floor?

Lake 12,460 in Sequoia National Park, California

Freshwater Environments

Let's begin our tour of aquatic environments on Earth. There are two kinds of freshwater environments, standing-water environments and flowing-water environments. Lakes are the most common standing-water environments. Other standing-water environments are ponds and **vernal pools**. Rivers, streams, and creeks are flowing-water environments.

Lakes

Lakes are bodies of water surrounded by land. They are all over the world. Some lakes are low in valleys, like the Finger Lakes in New York State. Others are high in mountains, like Lake 12,460 in Sequoia National Park. It's high in the Sierra Nevada range in California.

Less than 1 percent of Earth's water is in freshwater lakes. Although lakes are called standing-water environments, the water in lakes is always moving. Water moves from one part of the lake to another. Streams flowing into the lake move the water. When the weather gets cold, water near the surface gets cold and sinks toward the bottom. Moving water carries oxygen and nutrients to other parts of the lake.

27

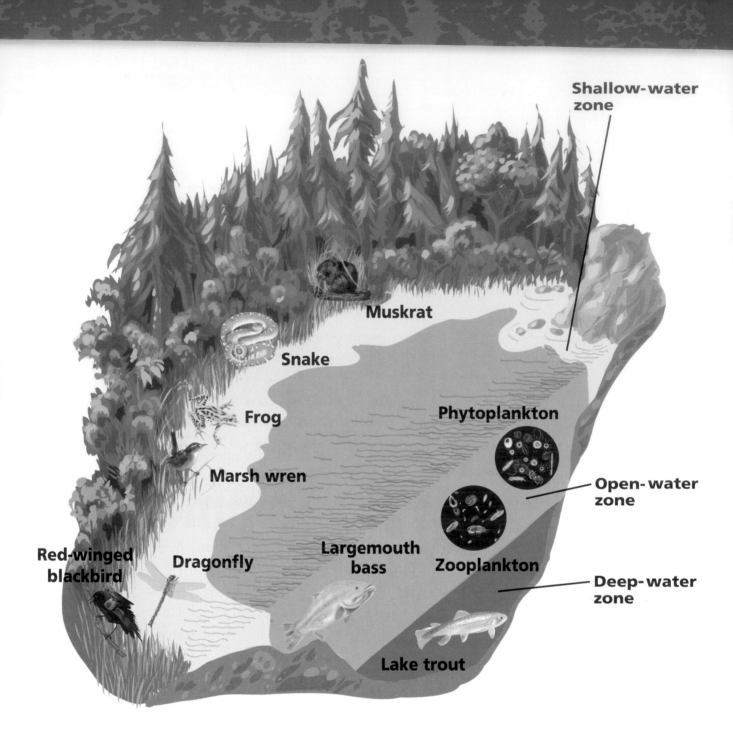

Shallow-water zone

Muskrat

Snake

Frog

Marsh wren

Phytoplankton

Open-water zone

Red-winged blackbird

Dragonfly

Largemouth bass

Zooplankton

Deep-water zone

Lake trout

Large lakes often have three different zones. These are the shallow-water zone, the open-water zone, and the deep-water zone. Each zone provides a different environment.

The shallow-water zone is near the shore. The water is shallow enough for sunlight to reach the lake bottom. The shallow-water zone often has rooted plants, such as water lilies, growing in the muddy bottom. Floating plants and **algae** may cover the surface of the water. Insect larvae swim around the plants and burrow into the mud. Insect larvae are food for larger animals, such as fish and frogs. Ducks, other birds, and small mammals live in the shallow-water zone.

The open-water zone is farther out in the lake. The water is deeper. There are no rooted plants. Sunlight is bright near the surface of the open-water zone. The light is dim in the deeper water.

Two important kinds of microscopic organisms live in this open-water zone. **Phytoplankton** are tiny plantlike organisms. They are the "grass" of the lake environment. They are eaten by **zooplankton**.

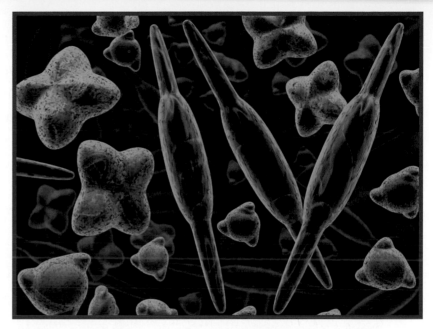

Phytoplankton are tiny plantlike organisms.

Zooplankton are microscopic animals in the lake environment. The zooplankton are food for insects and baby fish in the lake. Larger fish, such as trout and largemouth bass, live in the open- and deep-water zones. Birds, such as ducks, gulls, and grebes, also live in the open-water zone.

In big lakes, there is a deep-water zone. This zone is dark, and the water is cold all the time. Water from higher levels in the lake doesn't mix with the deep water, so there is less oxygen in the deep-water zone. Only animals that need little oxygen and light live at the bottom of the lake. These animals include some insects and lake trout.

The World's Five Largest Freshwater Lakes
(based on surface area)

Lake	Location	Area in square kilometers
Lake Superior	North America	82,103
Lake Victoria	Africa	69,484
Lake Huron	North America	59,596
Lake Michigan	North America	57,757
Lake Tanganyika	Africa	32,893

Ponds

A pond is a small, shallow body of water. Sunlight reaches the bottom of the pond. Plants and algae may cover the entire surface of the pond. Plants may be rooted or floating. Ponds also have large numbers of phytoplankton and zooplankton. Some animals that live in ponds are fish, birds, crayfish, frogs, snails, scuds, insects, turtles, and worms.

Plants and algae can grow over the surface of a pond.

Lake Baikal

Lake Baikal in Russia is the world's deepest freshwater lake. It is also the largest freshwater lake by volume. Twenty percent of the fresh water on Earth is in this one lake. It is 1,637 meters (m) deep and has a surface area of 31,468 square kilometers (km). The lake formed about 25 million years ago and is the oldest lake in the world. There are over 1,500 kinds of animals that live only in the Lake Baikal region.

Pollution from a nearby paper-making factory almost destroyed many of these animals in the 1950s and 1960s. Efforts to clean up the pollution have brought back much of the wildlife. However, Lake Baikal remains threatened by pollution.

Vernal Pools

A vernal pool is a shallow, temporary pond. *Vernal* means "spring." Vernal pools form when water collects in low places in the land. This happens during the rainy season or when snow melts in the spring. Vernal pools dry up during the dry season or the summer. When it is filled with water, the vernal pool is full of life. When it is dry, it looks like a mud flat. Vernal-pool plants and animals remain dormant during dry periods. When the pools fill with water again, the organisms reproduce and thrive. Salamanders, frogs, and many insects reproduce in vernal pools.

Rivers

Rivers are large bodies of moving fresh water. They usually flow into other rivers or into the ocean. Rivers often flow faster near their **source** in the mountains where the land is steep. Animals that live in the upper part of a river survive by being good swimmers or by holding tightly to rocks and twigs. Trout are strong swimmers. Insect larvae have hooks for holding on. As rivers flow toward the ocean, their currents may slow. Plants and animals in the lower parts of rivers are more like those that live in lakes. Smaller moving-water environments include creeks, brooks, and streams.

The World's Five Longest Rivers

River	Location	Length in kilometers
Nile	Africa	6,825
Amazon	South America	6,437
Yangtze	Asia	6,300
Mississippi/Missouri	North America	5,970
Yenisey	Asia	5,540

Thinking about Freshwater Environments

1. What living and nonliving factors define a lake's shallow-water zone?

2. What role do phytoplankton play in a freshwater environment?

3. How are lake and river organisms different?

What Is an Ecosystem?

An **ecosystem** is a **community** of organisms **interacting** with its nonliving environment. A terrarium is an ecosystem. When you put plants and a few small animals in a closed or semi-closed environment, the plants and animals are the living community. The soil, air, and water are the main nonliving factors in a terrarium environment.

An aquarium is also an ecosystem. Aquariums have a community of fish, snails, and water plants. Water is the main nonliving factor in an aquarium environment.

Terrariums and aquariums are two kinds of ecosystems. There are many more. When you go to the forest, you are visiting a natural ecosystem. Trees, grass, squirrels, birds, and insects are some of the organisms interacting with the nonliving environment in the forest ecosystem.

Ecologists are scientists who study ecosystems. They find out what kinds of plants, animals, and other organisms are living in an area. They also observe properties and measure the nonliving environmental factors in the area (the air temperature, soil, water, and light). Then they study how the community of organisms and the nonliving environmental factors interact.

A terrarium

An aquarium

32

Matter and Energy in an Ecosystem

Living organisms need food to survive. Food provides **matter** and **energy**. Matter is stuff. Everything that takes up space is matter. Air, water, rock, wood, metal, machines, buildings, and organisms are all matter.

Energy makes things happen. Energy makes it possible for organisms to grow and move. Organisms use energy to sense their environment and to reproduce.

Organisms get both matter and energy from food. But the way plants get the food they need for life is very different from the way animals get their food.

Plants *make* the food they need for life. Plants get the matter and energy they need to make food from air, water, and light in the environment.

Animals *cannot* make their own food as plants do. Animals get the food they need for life from other organisms. The way animals get food from other organisms is to eat them.

A forest ecosystem

Energy for most ecosystems comes from the Sun. Energy from the Sun is captured by plants' green leaves. Plants use water (H_2O), carbon dioxide (CO_2), and sunlight from the nonliving environment to make sugar. This process is called **photosynthesis**. Plants then use the sugar as food.

When animals eat plants, the energy of the sugar transfers to the animal. But even though animals get the energy from plants, it is really energy that came from the Sun. All the energy that makes living organisms move, grow, and reproduce in most ecosystems comes from the Sun.

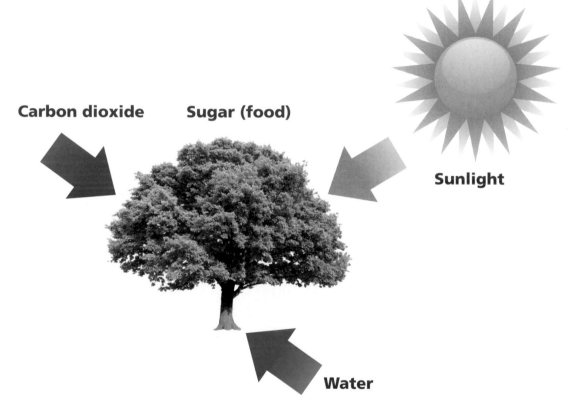

Carbon dioxide **Sugar (food)**

Sunlight

Water

Plants use water, carbon dioxide, and sunlight to make their own food.

Thinking about Ecosystems

1. How do plants and animals get the food they need to survive?

2. Explain how energy from the Sun helps animals survive.

3. What is an ecosystem?

Food Chains and Food Webs

In any ecosystem, a lot of eating is going on. Do you remember why? Eating is the way animals get the food they need to survive. What is it about food that makes life possible? Food is a source of matter and energy. The matter in food provides the raw materials an organism needs to grow and reproduce. Energy is like fuel that makes things happen.

One way to think about ecosystems is who eats whom. When you know how an organism gets its food, you can put it into a group. Let's look at the groups.

Producers

Some organisms don't eat anything. They don't have to because they make their own food. Organisms that make their own food are called **producers**. In terrestrial ecosystems, the most important producers are plants. Grasses, trees, and bushes are producers. In freshwater and ocean ecosystems, algae are the most important producers.

Algae are organisms that play an important role in aquatic ecosystems. Many algae are microscopic. Algae produce most of the food in freshwater and ocean ecosystems. They use water (H_2O), carbon dioxide (CO_2), and sunlight to make their own food, just like plants. Algae are the food source for many kinds of crustaceans, insects, fish, and worms. In your goldfish aquarium, you might have seen algae growing. Did the water turn green? Did a green layer form on the sides of the aquarium? If so, then you saw algae.

A type of freshwater algae called *Oedogonium*

But wait! If algae are **microorganisms**, how can you see them? When a few algae are in your aquarium, you won't see them because they are so small. But they start to reproduce. And after a week or two, the population of algae will be in the billions! That's what you see. Any one of those microorganisms by itself is much too small to see. You need a microscope to see just one. But huge numbers of them can affect the color and clarity of the water, making it look green and cloudy.

What happens to the algae? In a freshwater lake, insects and fish eat the algae. In the ocean, algae are food for baby clams, barnacles, corals, and thousands of young fish, crabs, and snails.

Producers use the food they make as a source of matter and energy. They don't eat other organisms for matter and energy. Any organism that makes its own food is a producer.

Algae on a pond

A ground squirrel

Caterpillars

A vulture

A snake

Consumers

Organisms that eat other organisms are **consumers**. Consumers can't make their own food. Consumers have to eat other organisms to get their matter and energy.

Some consumers eat plants and plant parts. Deer eat grass and leaves. Gophers eat roots. Squirrels eat grass, nuts, and berries. Caterpillars eat leaves. Animals that eat only plants to get their food are called **herbivores**.

Some animals don't eat plants. Snakes don't eat nuts and berries. Hawks don't eat grass. Spiders don't eat leaves. So how do they get their matter and energy? They eat other animals. Snakes and hawks eat gophers and squirrels. Spiders eat insects. Animals that eat other animals are called **carnivores**.

Some consumers, like humans, bears, raccoons, robins, and crayfish, eat both plants and animals. They are called **omnivores**.

Scavengers are consumers that eat dead organisms. Some scavengers, like vultures, eat only dead animals. Others, like isopods and termites, eat dead leaves and wood. Coyotes, rats, ants, and earthworms will eat just about anything that is dead.

Decomposers

There is a hidden world in every ecosystem. Millions of insects and invisible microorganisms use the last bits of dead plants and animals for food. They can be thought of as the cleanup crew. These organisms are called **decomposers**.

Decompose means "to break into parts." Insects, such as ants and termites, break down dead plants and animals into tiny pieces. Then the decomposers, the **bacteria** and fungi, take over. Bacteria and fungi break down dead plant and animal matter into simple chemicals

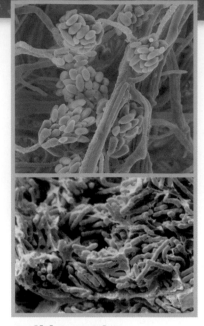

Soil bacteria

(nutrients). The simple chemicals are returned to the environment. When decomposers are done with a dead organism, there is no energy to transfer, and there is no longer any food value. The simple chemicals are the raw materials used by producers to make more food. Decomposers are the ecosystem's recyclers of matter.

Bacteria are the smallest organisms in the world. They are found in all environments. Bacteria play a very important role in every ecosystem. Bacteria decompose dead matter and waste. After bacteria finish their work, there is no energy to transfer from the matter, and the raw materials are returned to the environment. Some bacteria can cause disease, but most bacteria have important roles in ecosystems.

Fungi are important decomposers, too. They come in different shapes, sizes, and types. We know them as molds, mildew, and mushrooms. Like bacteria, fungi can live everywhere. They can live in both terrestrial and aquatic ecosystems. They are in the soil, in your home, on plants and animals, and even on you. A spoonful of soil might contain 120,000 fungi. Some are harmful to living plants and animals. But most fungi are important in recycling dead matter for raw materials in the environment.

Mushrooms are fungi.

Food Chains

When a spider eats a fly, the matter and energy in the fly go to the spider. This feeding relationship can be shown with an arrow. The arrow always points in the direction that the matter and energy flow.

fly   **spider**

If a praying mantis eats a spider, the matter and energy in the spider go to the praying mantis.

fly   **spider**   **praying mantis**

It's possible in a woodland ecosystem for a blue jay to eat the praying mantis, a weasel to eat the blue jay, and a hawk to eat the weasel. Matter and energy pass from one organism to the next when they are eaten. This is called a **food chain**. And at the beginning of the food chain is a producer. Energy for producers comes from the Sun.

In this case, the producer is a fruit from a tree, a plum. You can draw arrows from one organism to the next to describe a food chain. The arrows show the direction of energy flow. They point from the organism that is eaten to the organism that eats it.

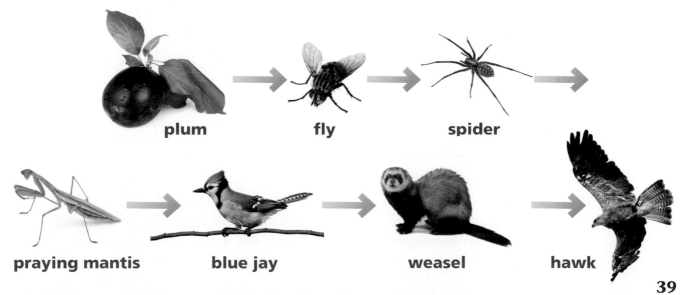

plum   **fly**   **spider**

praying mantis   **blue jay**   **weasel**   **hawk**

39

Another example of a food chain might have grass as the producer. A chipmunk eats the grass seed. A hawk eats the chipmunk. Bacteria decompose any dead organisms or uneaten parts. You can always draw arrows from dead organisms to the decomposers.

A simple food chain

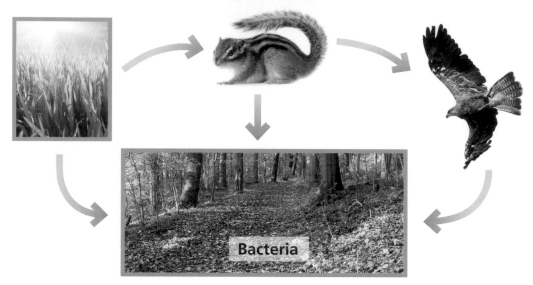

Bacteria

Food Webs

There are many feeding relationships in an ecosystem. If you draw *all* the arrows that show who eats whom, you have a **food web**, not a food chain. The food web for a freshwater river might look like this.

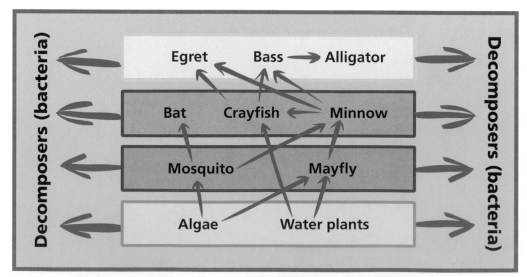

This is an example of a food web for a freshwater river. Bacteria decompose all the organisms when they die.

Locate the crayfish in the example of a food web. Crayfish are food for both egrets and bass. If the river has a lot of crayfish, egrets and bass will both have plenty to eat. But if there are few crayfish, the egrets and bass will have to **compete** with each other for food.

The animal that can get more food is the one that is more likely to survive. In this river ecosystem, egrets and bass compete for crayfish. Are there other competitions for food in the ecosystem?

Organisms in ecosystems depend on one another for the food they need to survive. Herbivores depend on producers to make food. Carnivores depend on consumers for food. Omnivores depend on both producers and consumers for food. Decomposers depend on dead organisms and waste for food. And producers depend on decomposers for raw materials to make food. In a healthy ecosystem, some organisms will be eaten so that other organisms will survive.

Egrets eat crayfish.

Thinking about Food Chains and Food Webs

1. What is food? Why is it important?

2. Do plants need food? Why or why not?

3. What is the role of producers in an ecosystem?

4. Look at the food web for a freshwater river. Give three examples of animals that compete for a food source.

5. What is the role of decomposers in an ecosystem?

6. How might a forest fire affect the food web in a forest?

Human Activities and Aquatic Ecosystems

The Lake Erie Story

Lake Erie is one of the five Great Lakes on the border between the United States and Canada. The other four lakes are Lake Superior, Lake Michigan, Lake Huron, and Lake Ontario. Together they form the largest freshwater body in the world. In fact, 95 percent of our nation's fresh water is in the Great Lakes.

Humans caused an ecological problem in Lake Erie. For years, sewage, farm runoff, and industrial waste were dumped and washed into Lake Erie. By the 1960s, the lake had become very **polluted**. The environment had changed. Too much algae grew in the water. The amount of oxygen in the lake dropped. This made it difficult for fish to survive. There were large areas of the lake bottom where life did not exist. The lake's aquatic organisms were dying. The ecosystem was in serious trouble.

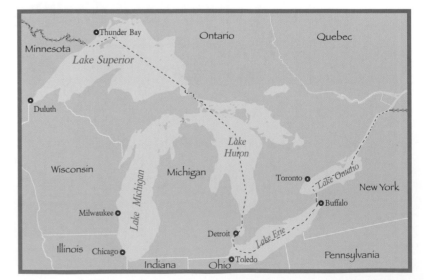

A coal power plant at Lake Erie, New York

In the 1970s, people started to save Lake Erie. The United States and Canada cleaned it up. Together, the two countries spent millions of dollars to develop a plan to save Lake Erie. Here is part of the plan.

- Build new and better sewage treatment plants.
- Reduce the use of detergents containing phosphates. Phosphates act as **fertilizer** for algae.
- Manage the use of fertilizers and **pesticides** on farms.
- Stop industries from dumping waste into the lake.

The efforts of both governments and the people living and working around Lake Erie paid off. The lake began to recover. After many years, the lake is much safer and healthier. The lake environment now supports many fish and other animals.

Lake Erie still has problems. More than 300 human-made chemicals are in Lake Erie. Some are poisonous. The United States and Canada continue to study ways to improve the water quality. Much has been done to help this important aquatic ecosystem. But the fight to save Lake Erie is not over.

The Lake Erie story is an important one in US history, but there are similar stories to be told everywhere. There are lakes, rivers, streams, and creeks all over the country that are suffering from human pollution. Let's dig deeper into the sources of this pollution.

A nuclear power plant on Lake Erie

Sources of Water Pollution

- **Farming** Pesticides are poisons used to kill pests on crops. Fertilizers are used to make crops grow faster and larger. When farmers use too much pesticide and fertilizer, the extra can wash into rivers and lakes. Pesticides kill plants and animals in the water. Fertilizers cause too much growth of aquatic plants and algae. This upsets the balance in aquatic ecosystems.

- **Sewage** Human sewage and waste from farm animals can also get into aquatic systems. These act like fertilizers, causing aquatic plants and algae to grow. Sewage can also carry microorganisms that cause diseases in humans.

- **Sediment** Runoff is water that flows over the land and then into large bodies of water. Runoff can carry soil and chemicals from mines, fields, forests, and cities. These materials settle to the bottom of lakes as sediments. Sediments can bury aquatic plants and animals. This burial can damage the environment and the organisms that live there.

- **Acid** Industrial gases from smokestacks enter the air. The gases form acid in clouds. Acid rain falls from these clouds and changes the acid levels in aquatic ecosystems. Many aquatic plants and animals are sensitive to acid. When the acid level changes, some organisms continue to survive, some survive poorly, and some cannot survive at all.

- **Petroleum** Oil spills and runoff from city streets can put oil and other **petroleum** products into lakes. Oil is harmful to animals that live on the lake surface, like waterbirds. It also harms organisms that live on the lakeshore, like snails, insects, and crayfish.

World Lakes Getting Warmer

In addition to water pollution, there is another concern for the health of freshwater environments. In 2010, NASA scientists published a study comparing surface temperatures of 167 lakes around the world. The scientists used satellite data to determine the water's surface temperature. The data were collected over a 25-year period, starting in 1985. The scientists used only summertime measurements to avoid winter weather and clouds that would block the satellite's view. They used only nighttime measurements to avoid effects of reflected sunlight. They selected lakes that were large, at least 500 square kilometers (km). All the lakes were inland, far from coastal shores. The measurements were taken in the middle of the lakes to avoid interference from the land.

What did they find out? The temperatures rose significantly in these lakes over the 25 years. The average increase in surface temperature was 0.05 degrees Celsius (°C) for every 10 years. In some lakes, the increase was as high as 0.10°C. That might not seem like a lot, but a small temperature change can have a big effect on a lake ecosystem.

Think about a lake freezing during the winter. If the lake freezes later each year and warms up earlier each year, that can have a big impact on the organisms that live in the lake. For example, nonnative **species**, either a plant or a fish, could become established in the warmer lake. A nonnative species might use the food resources in the lake and make it more difficult for native species to compete for food.

Human activities impact Earth's aquatic systems.

This is one of the important studies looking at **climate** change worldwide and its impact on freshwater lakes. Other studies have documented that the temperature of Earth's land and water surfaces are increasing in large part because of human activities.

Thinking about Aquatic Systems

1. What organism causes most of the pollution in Lake Erie? Give examples of why you think so.

2. What is the effect of climate change on large freshwater lakes?

A terrestrial ecosystem

Comparing Aquatic and Terrestrial Ecosystems

Aquatic and terrestrial ecosystems are very different. But they are the same in some ways. Let's compare.

The nonliving factors of the two environments are different. Aquatic ecosystems are in water. Terrestrial ecosystems are on land. The temperature in an aquatic ecosystem changes slowly. The temperature in a terrestrial ecosystem can change rapidly over a short period of time. The amount of water in an aquatic ecosystem is predictable. Water in a terrestrial ecosystem can vary widely.

The organisms are different in the two ecosystems. Most aquatic organisms can live only in water. If they were moved to a terrestrial ecosystem, they would die. The same is true for terrestrial organisms moved into aquatic ecosystems.

An aquatic ecosystem

A heron is a consumer of crayfish in an aquatic ecosystem.

A fox is a consumer of mice in a terrestrial ecosystem.

Both ecosystems, however, are organized in similar ways. The organisms in aquatic and terrestrial ecosystems all need matter and energy to stay alive.

- Both ecosystems obtain energy from the Sun and matter from the environment.
- Both have food chains and food webs.
- Both have consumers that depend on producers to make food.
- Both have decomposers that break down dead organisms and recycle the raw materials (nutrients).
- Herbivores, carnivores, omnivores, and scavengers live in both ecosystems.

In both ecosystems, organisms compete for the resources they need to survive. Plants compete for light. Animals compete for food. Organisms need space and shelter from predators and changes in the nonliving environment. The organism that outcompetes the others is the organism that will survive.

Animal Sensory Systems

How do you get the information you need to survive in your environment? You get information through your five **senses**. You use your sense of hearing, touch or feel, sight, smell, and taste.

Your senses make you aware of suitable food through smell and taste. They make you aware of things far away through sight and hearing. And they help you sense things that are close, through smell and touch or feel. The sense of touch or feel has many dimensions. Through touch, you can detect many kinds of input to your skin. Human touch can detect pressure, heat, cold, pain, tickle, itch, and textures such as smooth, rough, slippery, and sharp.

Sensory receptors on your body get information from the environment. Some of this information travels to your brain, which processes it. Then it sends information to your body to take action.

Animals use these same senses, or **variations** of these senses, to get the information they need to survive in their environments. Some animals use senses that are beyond the reach of humans.

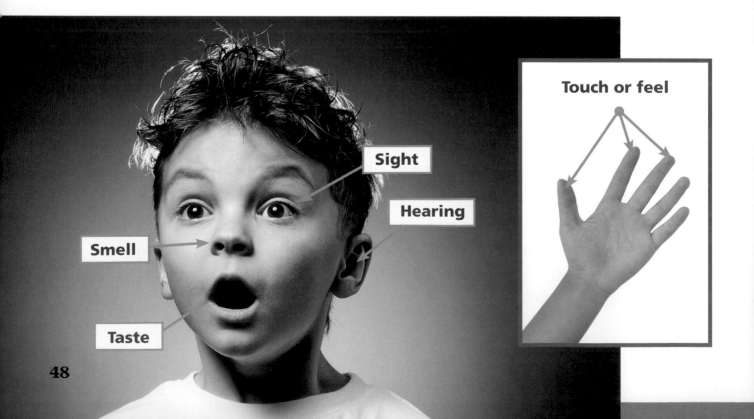

Sight

Hearing

Smell

Taste

Touch or feel

Rattlesnakes are members of a family of snakes called pitvipers. Pitvipers have a sensory receptor on their face that detects heat. With this structure, rattlesnakes locate **prey** such as mice and rats. These small animals give off body heat. The snakes can sense the heat. Put a blindfold on a rattlesnake, and it can still sense and capture prey. A rattlesnake can find a meal even in complete darkness.

American pitvipers include 16 kinds of rattlesnake, the copperhead, and the cottonmouth. Other kinds of pitvipers live in Central and South America.

A western diamondback rattlesnake

Close-up of a rattlesnake's head showing heat-sensing pit

pit

Honeybees must find flowers with sweet nectar and pollen to survive. Their eyes are part of a sensory system that provides important information about flowers. The light that we see reflected by flowers provides information for bees. But honeybee eyes also detect ultraviolet light. The light reflected from a flower looks quite different to a bee than it does to us. Many flowers appear to have a bull's-eye design. That design helps the bee locate nectar quickly. A white circle with a red center tells the bee to go to that flower and get the nectar and pollen.

This flower appears uniformly bright yellow to a human eye, but white with a vivid red center to a honeybee.

This flower also looks yellow to the human eye. But it appears white with a bright red center to a honeybee.

These small fish are swimming in a group called a school. Schooling fish can swim close to one another and change direction all together. They do this without colliding because they have a sensory system called a lateral line. The lateral line looks like a line of little dots running from the head to the tail.

Lateral line

The lateral line detects changes in water pressure along the fish's side. Sensing changes in pressure allows thousands of fish to swim together safely in a single school.

Pressure information can be helpful to large predator fish and small prey fish. Changes in pressure can alert predator fish to potential prey swimming nearby. This same information can signal prey fish that a predator is nearby.

The lateral-line sense is most like hearing in humans. However, humans cannot detect pressure changes in the same way.

Humans have no sensory system to detect magnetism. But evidence shows that some animals on Earth can sense and use magnetic information.

Some fish and birds can sense Earth's magnetic field. These animals might use this information to find their way when traveling long distances.

People carry homing pigeons in dark boxes to unfamiliar locations far from their home. When the pigeons are released, they fly up into the air. They circle a little while and then start flying directly toward home. Scientists have discovered that the pigeons have a space in their beaks that contains iron particles. The space and iron particles act like a compass.

Homing pigeons

Salmon may also have a magnetic sensory system. Scientists think that salmons' magnetic sensory system helps guide the fish back to the river where they were born. Salmon live as adults in the ocean for several years. Then, they swim up river to their birthplace to mate and reproduce. Their sense of smell also helps them find their birthplace.

Salmon

Scientists also suspect that sea turtles use magnetic information to guide them back to the beach where they were born. Adult sea turtles return to their birthplace after several years of life in the ocean.

A green sea turtle

Bats use hypersound to navigate and find insects to eat.

Bats are excellent navigators in the dark. They can fly through caves and catch insects during the darkest nights. Is this because their eyes are different than ours? No, bats do not use eyesight to navigate in the dark. They use **hypersound**.

Hypersound has very high frequencies that are too high for human ears to detect. But bat ears can hear hypersound. Bats produce hypersounds that bounce off objects such as insects and the walls of caves. Their large ears then detect the reflected sounds. This high-frequency hearing lets a bat know the exact locations of objects in its environment.

Other animals, like the cricket, also use hypersound to communicate. The familiar chirping of crickets is called **stridulation**. The cricket's stridulations are made of different sounds. They are something like the many sounds of an orchestra. Some sounds in a stridulation are within human hearing range. But much of a stridulation is so high that it can be detected only by another cricket.

Crickets can produce several kinds of stridulation by rubbing their wings together. Male crickets make a loud, harsh stridulation to claim a territory. This sound tells other males to stay away. They use a softer, sweeter, stridulation to attract a female. This sound allows cricket pairs to find each other.

A cricket

Very large animals, like elephants and whales, can produce and hear **ultrasound**. These ultrasounds are so low that most sensory receptors cannot detect them. If you were standing near an elephant producing ultrasound, you might feel a rumbling. It might feel like the shaking of the ground when a large truck drives by. But your ears would not hear the elephant sound. Low ultrasounds can travel long distances through air or water. These sounds allow elephants and whales to communicate at a distance with others of their kind.

Sharks have an awesome set of sensory systems. They have good eyesight at close distances. Like cats, sharks have excellent night vision. Like all fish, sharks have a lateral line to detect changes in water movement and pressure.

The shark's sense of smell is also exceptional. With two large nostrils on the sides of its head, a shark can sense a smell and tell its direction. Sharks can follow the smell of a tiny amount of blood in the water to locate and eat an injured fish.

In addition, the shark has a structure called the ampulla of Lorenzini. This structure allows the shark to detect electric signals in its environment. Injured fish produce electric signals. The electricity is not like the electric shock you could get from an electric circuit. The electricity produced by an injured fish is very weak. Sharks use the electricity sensors in their snout to find the injured fish.

Awareness of the environment and quick responses are essential for survival. Animals have different kinds of sensors that pick up clues from the environment, both far and near. Each animal has different systems for interpreting the information and responding. A human's brain processes sensory input and decides what action to take.

nostrils

Tiny dots on the shark's snout (red circle) hold the electricity-sensing ampullae of Lorenzini.

A marbled murrelet

Saving Murrelets through Mimicry

The marbled murrelet is a robin-sized seabird. It spends most of its life in the open ocean from Alaska south to central California. Marbled murrelets eat fish and other small ocean animals. They have streamlined bodies for diving, and strong, sharp beaks for capturing food. The murrelets' structures and behavior suit them for life on the open ocean.

Like all birds, murrelets must leave the water when it is time to breed and to raise their young. Most seabirds build nests on cliff ledges. But marbled murrelets raise their young in trees in coastal forests. Once a year, a breeding pair of murrelets will fly to the top of a tall redwood or other large tree. They nest in the same tree each year. There, the female lays a single egg on a moss-covered limb. The parents take turns sitting on the speckled blue-green egg. For 2 months, the parents fly back and forth from the nesting tree to the ocean to get food. Each trip might be 50 kilometers (km).

Murrelet females lay only one egg a year. The loss of an egg means that a murrelet pair will not produce any offspring that year.

Steller's jays are major predators of murrelet eggs.

Wildlife biologists are concerned about the survival of marbled murrelets. Their survival mainly depends on two factors. The murrelets need large mature trees near the ocean for nesting. And they need safety for their eggs and chicks. Predators often eat unguarded murrelet egg. One frequent egg predator is the Steller's jay. When a jay spots an unguarded murrelet egg, it swoops into the nest. The jay uses its sharp beak to break the egg and eat it.

Biologists are testing ways to change the jay's behavior so they won't destroy the murrelet eggs. One experiment involves painting small chicken eggs to look like murrelet eggs. The biologist injects a drug into the murrelet-egg mimic. The drug makes the jays vomit, but doesn't do any lasting harm to them. The biologist then puts the mimic eggs in the trees near where the murrelets nest. The mimic eggs are low in the trees, and the real eggs are high in the trees. When a jay eats a mimic egg, it immediately vomits.

In the experiment, jays that ate mimic eggs learned to avoid all speckled blue-green eggs. These smart birds might even be able to warn other jays not to eat those eggs.

Jays are corvids, a particularly intelligent family of birds. The corvid family includes jays, crows, and ravens. There is research that suggests that corvids are able to communicate what they learn to others of their kind. A jay that has tried the mimic egg and gotten sick, may communicate the message to others that says, "Don't eat those blue eggs with the brown marks; you'll get sick!"

Biologists are trying to decrease the number of jays living near murrelet nests. The population of jays has increased because human visitors to the forest are feeding them. Humans feed the jays for fun or accidentally by leaving picnic crumbs behind. More jays stay in the area because they can easily find food. Biologists are teaching people to not feed the jays. Jays will be less likely to stay in the area and less likely to find and eat the murrelet eggs.

In a 2010 study, biologists have estimated that up to 80 percent of the murrelet eggs along the central California coast were lost to predators each year. They predicted that this murrelet population could disappear within 100 years. But this might not happen now. By placing the mimic eggs in the murrelet habitat, keeping picnic areas clean, and preserving the largest forest trees, people are helping murrelet eggs survive. This is good news for the chances of the continuation of the central California population of marbled murrelets.

These coast redwoods in which murrelets nest are the tallest trees in the world.

Brine Shrimp

Brine shrimp (*Artemia*) are small crustaceans. They live only in salty aquatic environments, such as salt lakes and brine ponds. Unlike their relatives, crabs and lobsters, brine shrimp cannot live in the open ocean. But brine shrimp can live in environments where most other organisms cannot.

An adult brine shrimp is about 1 centimeter (cm) long. Its partly transparent (clear) body is divided into segments. Between 11 and 19 of the brine shrimp's segments have legs. The legs are used for swimming and feeding. As it swims, it pulls nearby microscopic bits of food into its mouth.

Salt is an important nonliving factor in the environment of brine shrimp. Brine shrimp can live in a **range** of salt **concentrations**. The brine shrimp living in Mono Lake thrive when the salt concentration in the environment is 80 parts salt per 1,000 parts water. If the salt concentration drops below 60 parts salt, the brine shrimp will survive, but not as well. If the salt concentration falls below 20 parts per 1,000, or goes over 100 parts per 1,000, the brine shrimp will not survive.

Brine shrimp eggs can survive during long periods of dryness. They can stay dormant for up to 3 years before hatching. Live adult brine shrimp are sold as food for larger fish. Brine shrimp are sometimes sold as pets called sea monkeys.

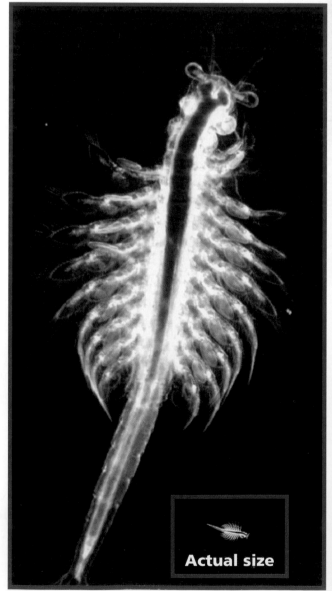

Actual size

Magnified view of a brine shrimp

A view of Mono Lake

The Mono Lake Story

Mono Lake lies at the edge of the Great Basin in northeastern California. It is at least 760,000 years old. That makes it the oldest lake in America. Because it lies between the Sierra Nevada and the desert, Mono Lake is blistering hot in the summer and freezing cold in the winter. Mono Lake's water is salty, even saltier than the ocean.

At first glance, Mono Lake looks lifeless. But it isn't. What looks like a lifeless lake is a rare and important ecosystem.

The story of the Mono Lake ecosystem starts with the lake water itself. Because it is so salty, no plants or common lake animals can live there. There are no fish, frogs, or mosquitoes in the water. But two kinds of algae thrive in the salty water. They are tiny floating algae and bottom algae. Floating algae drift around in the lake, and bottom algae grow on the lake bottom. These algae are the only producers in the Mono Lake ecosystem.

Pink clouds of brine shrimp in Mono Lake

A close-up of a brine shrimp

Tufa towers made of calcium carbonate form in Mono Lake.

Brine Shrimp

The most important animal in Mono Lake is the brine shrimp. Over the winter, the bottom of the lake is covered with billions of brine shrimp eggs. In late spring, the water starts to warm. The eggs start to hatch. The tiny brine shrimp are no larger than the period at the end of this sentence. The shrimp eat the floating algae and grow. In a few weeks, they are full-sized adults. They start to reproduce. By early summer, there are trillions of brine shrimp in the lake. Mobs of several million shrimp form pink clouds all over the lake.

Many birds **migrate** from winter feeding grounds to spring nesting grounds. Mono Lake plays an important role in the survival of several kinds of birds. About 50,000 California gulls migrate from the ocean to Mono Lake to breed. The gulls make nests on the two islands in Mono Lake and feed on the brine shrimp. When their eggs hatch, the gull parents catch brine shrimp to feed their chicks. By the middle of the summer, the chicks can fly. They follow their parents over the mountains to the ocean.

There is still more happening at Mono Lake. Small shorebirds called phalaropes and waterbirds called eared grebes also stop at Mono Lake to eat and rest. They reproduce in Canada and then fly farther south for the winter. Without Mono Lake as a place to rest and feed, they would not be able to finish their migration.

An amazing 150,000 phalaropes and between 1 and 2 million grebes come to Mono Lake during the summer. By midsummer, there is not much algae remaining in the lake. The brine shrimp have eaten most of the algae. There are trillions of brine shrimp in the lake. The phalaropes and grebes eat and eat and eat. By the time the birds are ready to continue their migration, only a few billion brine shrimp are left in the lake.

As the water cools in the fall, the last brine shrimp females lay eggs. These eggs don't hatch. They settle to the bottom of the lake. With the brine shrimp gone for the time being, the floating algae reproduce in huge numbers. The brine shrimp eggs lay dormant until the next spring. When the water warms up again, the eggs hatch. The new shrimp begin eating the algae, and the whole cycle happens again.

A Wilson's phalarope

An eared grebe

California gulls at Mono Lake

Brine flies along the water's edge

Close-up of brine flies

California gulls feed on brine flies.

Brine Flies

While the brine shrimp are eating the floating algae, the larvae of the brine fly are eating the bottom algae. When the larvae are full grown, they come to the lake's surface and pupate. In a few days, the **pupae** open and the adult flies come out. There are billions of them on the shore of Mono Lake.

The brine flies are another food source for the phalaropes, gulls, and grebes. In the past, brine flies were food for the native people living in the Mono basin. The word *mono* might be a Native American word that means "fly eater." The Native American people who lived in the basin would gather millions of the pupae, dry them in the sunshine, and store them for winter food. The pupae have a lot of protein and fat, and they were easy to harvest.

Humans and the Mono Lake Ecosystem

For thousands of years, Mono Lake was where the California gulls came to reproduce. In 1941, things began to change.

The Los Angeles Department of Water and Power began taking water from four of the seven streams that flow into Mono Lake. The water was sent 560 kilometers (km) away for the people in Los Angeles. The result was an ecological disaster for Mono Lake. The lake shrank to half its size. The salt concentration in the lake water doubled.

In 1982, the salt concentration in Mono Lake was making it hard for algae and brine flies to survive. The brine shrimp were in trouble, too. These changes affected the California gulls. They didn't have as much food to eat.

The lower water level created a land bridge between shore and one of the California gull nesting islands. Foxes and coyotes could walk to the island and eat the eggs and chicks. Because of the predators and lack of food, the California gulls did not raise any chicks that year.

Mono Lake Makes a Comeback

In 1978, a young man named David Gaines (1947–1988) became concerned about the poor environmental condition of Mono Lake. He formed an action group called the Mono Lake Committee. Under his leadership, the committee worked with government agencies, environmental groups, and the Los Angeles water department to solve the problem.

In 1994, a decision was finally reached. Less water would go to Los Angeles. This would allow the water level in Mono Lake to slowly rise.

Today the water in the lake has returned to a good level. The salt concentration has gone down. The land bridge to the island is again underwater. The brine shrimp and brine flies are thriving. The California gulls are raising chicks. The Mono Lake story shows that people can take positive action to restore the environment and save important ecosystems.

David Gaines

63

Mono Lake Food Web

Wilson's phalarope eats brine flies and brine shrimp.

The red-necked phalarope, Wilson's phalarope, and eared grebe stop at Mono Lake to eat and rest during migration. But they do not nest there.

The eared grebe is a diving bird. It eats brine flies and brine shrimp.

The coyote eats chicks and eggs.

The red-necked phalarope eats brine flies and brine shrimp.

The California gull nests on islands and eats brine shrimp and brine flies.

Brine shrimp live in all areas of open water.

Brine flies live in and near shallow water.

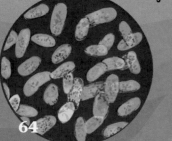

Floating algae

Bottom algae

Bacteria are decomposers.

Reviewing the Mono Lake Ecosystem

Mono Lake is a very salty ecosystem. Most aquatic organisms cannot survive in Mono Lake because of the high salt concentration. But a few kinds of algae thrive in the lake. These algae are the producers that support the Mono Lake ecosystem.

Some algae are small and float around in the lake. One individual alga is too small to see. That makes it a microorganism. But when countless billions of algae fill the lake, you can see them. The water turns green.

Like all ecosystems, Mono Lake has consumers. The two most important consumers are the brine shrimp and the brine flies. Millions of migrating birds stop at Mono Lake every year. The birds eat the brine shrimp and brine flies. The migrating birds rely on this food to survive.

Bacteria are the decomposers in Mono Lake. These microorganisms break down the dead algae, brine shrimp, brine flies, birds, and waste into simple chemicals. The simple chemicals recycle back into the ecosystem. These chemicals are the raw materials used by algae to make food the next year. Bacteria are an important part of the Mono Lake ecosystem.

One way to diagram the Mono Lake food web is shown on the previous page. Follow the arrows to see how energy moves in the ecosystem. A simple Mono Lake food web might look like this.

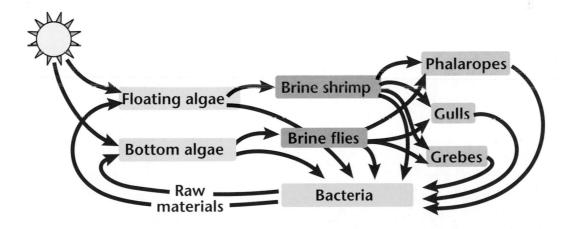

Thinking about Mono Lake

1. What is the main environmental factor that affects the health of the Mono Lake ecosystem? Explain your answer.

2. Why did the California gull chicks not survive at Mono Lake in 1982?

What Happens When Ecosystems Change?

Two things define an ecosystem. The first is the organisms. The second is the nonliving environment. The Sonoran Desert ecosystem of the southwest United States is partly defined by its common organisms. Cacti, lizards, and coyotes live in the desert. Spiders, insects, birds, scorpions, and snakes also live in the desert. The environment is hot and dry most of the time. The desert ecosystem thrives with these organisms and environmental conditions.

A desert ecosystem

A tropical marine ecosystem

The coral reef ecosystem of the Florida Keys National Marine Sanctuary is very different from the Sonoran Desert. The Keys Sanctuary is the only living coral barrier reef in North America. It is the third largest system of coral reefs in the world. The most important organisms in the reef ecosystem are coral and algae. Marine snails and parrot fish feed on the coral, and damselfish eat the algae. Fish of all sizes, shapes, and colors can be found swimming along the reef. Organisms such as spiny lobsters and long-spined black urchins come out from their hiding places and feed at night. Between the barrier reef and the shore are lagoons with calm water. On their sandy bottoms, turtle grass grows. Bottom dwellers, such as the southern stingray and gulf flounder, settle into the sand to feed. The environment is the tropical ocean.

Moving desert organisms to the ocean would be silly. Cacti and snakes from the desert would not survive in a marine environment. The environment is too different. Organisms have needs. Those needs are met only by the environment in which the organisms live. But what happens when the environment changes a little bit? Are organisms affected?

We can look at the Colorado River as an example. Before 1963, the river flowed through Grand Canyon. Each spring the melting snow upstream caused a flood in the river. The flood of cold water roared through the riverbed. It washed away the sandy beaches. It cleared out plants growing on the banks of the river.

During the summer, the river slowed. Sand settled on the edges of the river. The water warmed up. Plants began to grow on the new sandy beaches. The organisms in the river thrived in this environment.

The Colorado River ecosystem

The Glen Canyon Dam changed the Colorado River ecosystem.

In 1963, the Glen Canyon Dam was completed. The dam stops the flow of flood water each spring. Cold water flows through the bottom of the dam at a steady rate. There are no more floods. The water flowing in the river is cold all year long.

This change in the environment was small. But it affected the balance of organisms in the ecosystem. Trout need cold water to thrive. After the dam was completed, the trout population grew because the water was cold all year. The trout became predators of the young humpback chub living in the river. The humpback chub is a large fish that can live up to 40 years. It lives only in the Colorado River Basin. The humpback chub population got smaller because more trout were in the river.

Trout have a **range of tolerance** for temperature. If the water gets warmer than they can tolerate, or colder than they can tolerate, they will not survive. Before the dam was built, the warmer summer water was within the range of tolerance for trout, so they survived. But it was not the best temperature for the trout. After the dam was built, the colder water was the **optimum** (best) temperature for trout. As a result, the trout thrived, their population grew, and they ate more of the chub.

69

The willows and other plants that grew on the sandy beaches were no longer washed away in the spring floods. More plants grew in the changed environment. The plants stopped the sand from moving in the riverbed. As a result, the shallow water where the baby chubs and river snails grew disappeared. So the chub and snail populations became smaller.

In March 1967, the humpback chub was placed on the federal list of **endangered** species. The reasons were habitat loss because of changes in water temperature and water flow, and competition from the trout. Recently, the chub populations are starting to increase, but they are still endangered. Here are two possible reasons for the increase.

- People have removed many rainbow trout and brown trout from the river. These two fish prey on young fish and compete with humpback chub for food.
- Starting in 2003, the area experienced a drought. As a result of the drought, water temperature increased in the Colorado River. This warmer water helped native fish, including the chub.

This Colorado River story shows how living and nonliving environmental factors affect organisms. The environment controls which organisms will thrive, survive, and die. Small changes in temperature, water flow, or population size can change the balance of an ecosystem. The change favors some organisms and makes it harder for others.

A humpback chub

Thinking about Populations

How did the change of water temperature after the dam was completed affect the chub population? Why?

The Shrimp Club

After reading about the endangered humpback chub, our teacher told us about another endangered species, the Attwater's prairie chicken. In the early 1900s, millions of Attwater's prairie chickens lived in the huge grasslands in eastern Texas. Today there might be as few as 100 breeding pairs. What caused such a decline in the number of prairie chickens?

The prairie chicken completely depends on the grassland habitat for its survival. As human populations have grown, the grassland habitat has been lost. Some habitat has been developed for towns and roads. Some habitat has been changed into fields to grow crops. Habitat loss is the main reason that Attwater's prairie chicken is endangered. There is not enough good habitat left to provide grassland for food and space to raise young chicks.

In Texas, wildlife biologists and conservation groups are working together to restore grasslands and prevent existing grasslands from being destroyed. The future survival of Attwater's prairie chicken is looking better.

A prairie chicken

Our class was wondering if students could do anything to help an endangered animal. Our teacher told us about an endangered freshwater shrimp living in streams in Northern California. Students in Marin County, California, and their teacher, Laurette Rogers, started the Shrimp Club to see how they could help the endangered shrimp survive. We interviewed Ms. Rogers to find out more about the Shrimp Club.

Q: Are you still doing the Shrimp Club project?

A: I work doing ecological (habitat) restoration to help endangered Northern California species. My commitment to this work started when I was a teacher at the Brookside School in Marin County, California, in 1992. My fourth-grade class worked together to form the Shrimp Club. The Shrimp Club grew and changed into the Students and Teachers Restoring a Watershed (STRAW) Project, an organization that coordinates many restoration projects.

Q: What inspired you to do a restoration project to help an endangered species?

A: I didn't plan to do this work. In 1992, my class watched a video on endangered species. John, a student, asked, "What can we do to help?" I knew I couldn't ignore the question. The class felt we had to do something real, and we needed to help. I believe school should be about real life so that students can practice being leaders and then continue to be leaders as adults.

Laurette Rogers

1 cm

The endangered California freshwater shrimp

Q: What endangered species did your class decide they wanted to protect, and why?

A: The Adopt a Species organization in California suggested that we choose between three Northern California endangered species that live in our creeks: a trout, a salmon, and a freshwater shrimp. The entire class voted, and the shrimp won. We chose California freshwater shrimp *Syncaris pacifica*.

We found out that

- The shrimp have lived in local creeks since dinosaurs roamed Earth.
- Males are 3 centimeters (cm) long; females are 5 cm long.
- They feed on dead and decaying streamside vegetation.
- They are weak swimmers. They must cling to the roots of willows growing along the banks of the creeks.
- The shrimp live and reproduce in protected, shaded pools.
- Most of the creeks run through ranches where cows eat the streamside vegetation and tramp through the water, destroying the shrimp habitat.

The class was excited to study something obscure and local rather than something cute and fluffy living far away. The shrimp was something no one else was protecting. As we learned about the shrimp, we grew to love them. This was the start of what my students fondly started to refer to as the "Shrimp Club."

Q: How did your students decide what to do?

A: My 28 students and I brainstormed ideas of what to do. We figured out we had to restore the creeks to their condition prior to the introduction of cows into the watershed. We organized the class into committees that met regularly. Each committee prepared a goal sheet that included what they were working on and the current status of their part of the project.

The T-shirt committee's goal was to raise money. The newsletter committee developed a regular newsletter to share information with the school community. The public relations committee made presentations in front of companies and news media, and encouraged journalists to come to our restoration days to report on our work. One committee met with ranchers to plan where we could restore native vegetation by the streams on their land. The stamp committee made a huge rubber stamp that explained what the Shrimp Club was trying to do. They took home brand-new paper bags from grocery stores, stamped them, and returned them to the store for customers to read. A comic-strip committee formed because a student wanted to draw and write comics about the shrimp project.

★ HELP SAVE THE CALIFORNIA FRESHWATER SHRIMP ★

SAVE OUR SHRIMP!

Join the Shrimp Club

Brookside School

Ms. Rogers 4th Grade

> 66 I always thought the teacher made the path and we followed, but in this case the students made the path and the teacher followed. 99
>
> *Adam (age 10)*

74

Young native plants arrive in tubes for planting.

Q: What were some of the exciting things the students got to do?

A: In the spring of 1993, we did our first fieldwork on Stemple Creek within the Estero de San Antonio watershed. We planted native plants. We wrote letters to government officials, and students were invited to testify at hearings before local government agencies about the Endangered Species Act. We actually traveled to Washington, DC, to present the Shrimp Club to the National Fish and Wildlife Foundation (NFWF), to the Environmental Protection Agency (EPA), and to our representatives in Congress, Senator Barbara Boxer and Congresswoman Lynn Woolsey. It was a very big deal. We were even on the news. That was exciting!

The less glamorous work was exciting, too. We worked with professional stream-restoration scientists to remove invasive plant species and reintroduce a lot of native willow and oak trees. One day, we watched as a calf was born on a ranch. Students loved being around ranch animals and learning about ranch life. We saw snakes, hawks, and even a badger. Most importantly, we got to do real work to make a difference in our community.

Q: We heard you won an award. What did you win?

A: In 1993, we won the grand prize in the "A Pledge and a Promise" environmental awards, and the award included a check for $32,000! The whole community went nuts, and some were even dancing in the streets. But we were not doing the work to win awards. We were doing it to restore the habitat for our shrimp.

Q: It sure seems like everything went really well. Did you have problems?

A: We made a lot of mistakes. A lot of phone calls didn't produce results. Sometimes journalists would hang up on my students. One time the public relations committee arranged for a reporter to show up at a restoration field trip, but the students forgot to confirm it, and the reporter didn't show. We learned a lot from these experiences.

Despite all the challenges and obstacles, it felt like there were people waiting to help us. For example, a banker helped us set up a checking account with a 9-year-old student as a cosigner. The student got to write a check for $25,000. One tiny shrimp opened up the whole world.

Q: What exactly is a watershed? Why do we need habitat restoration in some areas?

A: When water falls on the land, it flows downhill, usually ending up in a creek or stream. A watershed is the area of land that drains into a particular body of water. The watershed is usually named for the body of water into which the water flows. The watershed starts at the highest point of land, and all the water that flows into a single creek defines a watershed.

We need habitat restoration because the way people have used land in the past has made the land less usable for the native plants and animals that lived there before people moved in. We try to make the land work for the native plants and animals again. We try to work hand in hand with nature and the people who live there now to make the ecosystem work well for the native organisms that live there. At one particular ranch, we observed 8 species of birds before restoration. Now we see 28 species.

Students plant native trees and shrubs.

Lots of equipment is needed to restore a watershed.

Q: What plants are used to restore a watershed?

A: We use native plants, the plants that evolved to live in that spot. Nonnative plants are plants that were brought in by people. Sometimes during restoration work, we're just removing invasive species, making space for the native species to grow and expand.

Q: Can one class make a difference?

A: Yes, if you have a good plan. If you want to be involved in a creek restoration project, you should have a professional creek restorationist to help you. A wildlife scientist will help make sure you do the right thing and not accidentally do more damage. If you follow a plan and get the cooperation of the community, then yes, absolutely! One class can make a big difference.

Q: Are you still involved in this work?

A: It is my whole life, and I'm in my 19th year of watershed-restoration work. I still go to most restoration fieldwork days.

Each year, STRAW brings about 100 classes to local creeks that we have identified as needing restoration work. I work with a team of professionals with incredible scientific expertise and other STRAW faculty who go into K–12 classrooms to give pre-field lessons. Students often say, "It was hard work, but it was fun."

Q: How did the restoration work that students did in 1992 impact their lives?

A: The main thing is the students feel empowered to make a difference in their lives. From my experience, I've seen that students gain self-knowledge about who they are in the world. It's important that students have their hands on the reins of their own learning, that you trust them, and give them the freedom to have creativity and ownership.

Students can make a difference.

Q: How is the California freshwater shrimp doing?

A: I was told it would take 50 years to see if the restoration would work, but the shrimp have already expanded their range because of the work students did. They have moved downstream holding on to the roots of the willows students planted.

STRAW continues as a project of Point Reyes Bird Observatory (PRBO) Conservation Science.

> **66** I think this project changed everything I thought we could do. I always thought kids meant nothing . . . [But] kids can make a difference. We are not just little dots. **99**
>
> *Megan (age 10)*

> **66** I have diary entries from your class. Wanting to be a marine biologist, I remember writing furiously to express how excited I was about all you taught.
>
> Thank you so much for teaching me more than a college degree and 4 years of 'real world' work experience could have taught me. I will carry memories from your class always. I hope that as I enter graduate school, I can experience that same thrill and excitement I had when I was 10. Your class has stuck with me more than any other over the years. **99**
>
> *Megan (age 28) in a letter to her fourth-grade teacher Laurette Rogers*

Variation and Selection

If every person in your school brought their pet dog to class, you would see a lot of variation. There might be a tiny Chihuahua or a big Bernese mountain dog. There might be short-legged dachshunds or tall golden retrievers. There might be thin dalmatians or round-looking bulldogs. Where did they all come from? Why do they look so different?

Evidence suggests that pet dogs evolved from the wolf. Scientists think that one kind of wolf might have been comfortable around humans as far back as 135,000 years ago. These wolves were not pets, but they lived near humans. Much later, about 2,500 years ago, humans used some kinds of dogs for hunting, protecting livestock, and carrying loads. Today, there are about 400 different breeds (kinds) of pet dogs. How did 400 breeds of dog come from the wolf?

Pet dogs evolved from the wolf.

Dogs show a lot of variation.

Selective Breeding

Suppose you wanted a hunting dog to chase badgers out of their burrows. The dog would need short legs. So you would find a dog with short legs and make sure it produced offspring. You would breed the offspring with another dog with very short legs. And when those pups grew up, you would again breed the short-legged offspring. In a few **generations**, you might have a lot of short-legged dogs. Some of these dogs would be able to go into burrows to catch a badger. This might have been how the short-legged dachshund breed came to be.

A dachshund has short legs.

Selective breeding is when humans select individual organisms to breed to produce offspring with certain traits. Humans decide which qualities they want in a dog. They find individual dogs in the population that have these traits. Then they breed them to produce offspring with those same traits.

Selective breeding has produced the 400 different breeds of dogs. It has also produced many breeds of horses, cats, dairy cows, wheat, peppers, tomatoes, and corn. The desirable traits can be very different. You might want a plant that grows fast or produces fruit with no seeds. Humans are good at using selective breeding to meet their needs.

Peppers and tomatoes have many different breeds.

There is more competition for food during the winter.

Young moose (offspring) inherit traits from their parents.

Natural Selection

In nature, the environment (not humans) selects the individual organisms that will produce offspring. Some individuals are selected to reproduce, and some are not.

Life is a struggle. Animals compete for food. They also compete to find a mate. Changes in the environment, such as climate change, can place pressure on organisms. Individuals in populations that are adapted to their environment have ways to respond to these pressures.

In all populations, there is variation from one individual to the next. Some individuals will be better at getting food. Others will be better at avoiding predators. Some will be better at dealing with cold weather. These variations are important when the environment changes.

A change in the environment can add more pressure to a population. The change might be a new predator, a forest fire, or less food to eat. Some individuals in the population will be able to live successfully in this changed environment. These successful individuals will reproduce. They will leave offspring with the traits that allow them to survive in the changing environment. The traits of the survivors are passed to the next generation as **inherited traits**. That's natural selection.

Darwin's Finches

In 1835, Charles Darwin visited the Galápagos Islands off the shore of Ecuador. He observed and described many kinds of birds with different beaks. Later, other scientists studied these same birds. These birds are known as Darwin's finches, and there are 13 of them. These finches gave scientists a new way of thinking about how changes occur in populations.

We now know that all 13 different finches on the Galápagos Islands evolved from one species of finch. That original species of finch arrived on the islands thousands of years ago. How did one species of finch evolve into 13 different species?

The Galápagos Islands

Darwin collected many finches, each with a different beak.

Scientists speculate that long ago, a big storm blew a small flock of mainland finches out to sea. The flock landed on some small, volcanic islands. The small population of finches was in a very different environment. The islands had food and places to nest, but they were not what the birds were used to.

There were seeds of several sizes from grasses, shrubs, and trees. In the population of mainland finches, there was variation in beak size. Individuals with larger, stronger beaks could crack large seeds. Individuals with smaller beaks could not. Finches with smaller beaks could more easily gather large numbers of small seeds. Individuals with large beaks could not eat the small seeds. Variations in beak size turned out to be helpful.

Medium male (left) and female (right) ground finches found on the Galápagos Islands

Finches that fed on large seeds mated and produced offspring with large, strong beaks. Their beaks allowed them to survive when large seeds were plentiful. Over time, the finches lived as separate groups because of the seeds they ate. Over many generations and many years, the large-beaked finches and the small-beaked finches were so different that they could no longer mate with each other to produce offspring. They had evolved into two new species. And the new species were different from the original mainland finches.

This same process of eating different foods based on beak size and shape produced other differences in the populations. Over time, all the differences created 13 different species. Each species was adapted to feed on a different food source.

What would happen to the large-beaked finches if the large seeds became scarce? They would have to find a new food source. It might be seeds of a different size or perhaps insects. Within the population of large-beaked finches would be individuals with smaller beaks. The small beaks would make it easier for those individuals to feed on a smaller food source. They would survive and reproduce. Their offspring would inherit the trait of smaller beaks. Individuals who survived would pass their traits to the next generation. The pressure of finding food would cause the population to shift to finches with smaller beaks. This would take many generations.

Darwin's Finches Today

Recently, the environment on the Galápagos Islands changed again. Fly larvae are like a **parasite** to baby finches. They burrow into a chick's body and make it sick. The finches now have to deal with this new pressure. It is not clear whether individuals in the finch populations have adaptations to protect themselves from the deadly fly larvae.

The struggle for survival goes on. Because the environment is always changing, the populations that survive and thrive are always changing. Sometimes the change in the environment is so fast or so extreme that no individuals in a population survive. Then the entire population dies or becomes extinct. Extinction is part of the process of natural selection.

A large female ground finch opening a seed

A common female cactus finch

Environmental Scientists

The behaviors of humans can change ecosystems throughout the world. Environmental scientists study changes to environments and ecosystems. They also research ways to save native habitats and prevent destruction. These environmental scientists have made a difference in environmental science.

Rachel Carson

In 1962, Rachel Carson's book *Silent Spring* was published. Many people think it is the most important book ever written about ecology. It changed how Americans think about their place in nature.

Carson (1907–1964) studied marine biology in college. In 1936, the US Bureau of Fisheries hired her. She was the first woman biologist ever hired by the bureau. During the 1940s and 1950s, she studied and wrote about life in the sea. She also started to see some disturbing things happening in the environment. Animals were dying. She figured out that the cause was pesticides.

Pesticides developed in the 1940s were used widely to kill mosquitoes, fruit flies, cabbage worms, and lots of other pests. Carson discovered that the poisons also killed all the other insects in the area where the pesticide was sprayed. The poison killed everything.

When pesticide spray drifted over a friend's bird sanctuary, the birds died. Carson was alarmed. She imagined a spring when no birds returned to the woods near her home. In 1957, she started writing a book. The horrible thought of spring without the music of singing birds became the name for her book, *Silent Spring*.

Carson writing in her study

The book got a lot of attention. The companies that made the pesticides tried to get the book banned. They attacked Carson and her scientific conclusions. She fought back and continued to present her evidence. Soon people started to listen, including President John F. Kennedy. President Kennedy ordered studies of a pesticide called DDT to see if Carson's ideas were right.

Carson warned us of a danger to Earth's ecosystems. It was up to others, including many environmental groups, to act on her warning. As a result of *Silent Spring,* DDT was banned in 1972. But Carson never saw that day. In 1964, she died of cancer.

Carson is best remembered for making people aware of the dangers of pesticides. People started thinking about what happens to plants and animals when we change the environment either by accident or on purpose.

Edward O. Wilson

As a child, Edward O. Wilson (1929–) loved nature. He loved exploring in the Alabama woods and nearby streams. When he was 9 years old, he read an article about ants. Their interesting behavior and the way they worked together fascinated him. Those Alabama ants started young Wilson's career as a scientist. When he was only 13 years old, he discovered the first fire ants in the United States.

Wilson had a hard time with math and some trouble reading. This did not stop him from studying science. He studied biology at the University of Alabama and Harvard University. Later, he became a professor at Harvard.

Wilson continued to study ants. In 1971, he wrote a book about the social behavior of ants. He described how ants and other animals communicate using chemicals called pheromones. In the book, he also compared the organization of ant colonies to human societies.

In 1992, Wilson wrote a book called *The Diversity of Life.* He wrote that human activities were destroying organisms worldwide. Wilson predicted that millions of kinds of plants and animals would become extinct by the middle of the 21st century. Since then, Wilson has been trying to find ways that humans can save the world's ecosystems.

In 2007, Wilson gave a speech challenging everyone to learn more about our biosphere. He asked for the development of an online database of what we know about organisms. His speech gave rise to the Encyclopedia of Life project (EOL), which provides global access to knowledge about life on Earth.

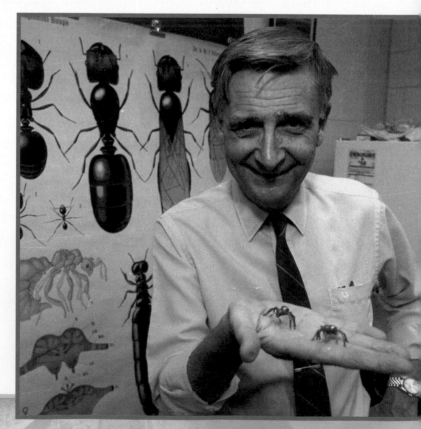

Tyrone B. Hayes

Tyrone B. Hayes (1967–) is a biology professor at the University of California at Berkeley. He studies frogs. The first frogs he saw were in a swamp near his home in South Carolina. Now Hayes studies frogs in Africa and North America.

Hayes found something strange happening to some of the frogs he studied. Frogs living in the wild were going through sex changes. The "male" frogs were making eggs just like the females.

Hayes and his team studied the frogs' environment. They found small amounts of a common **herbicide** in the water. An herbicide is a chemical used to kill plants. Hayes did more tests in the lab. The tests showed that the herbicide was causing the changes in the frogs.

The companies that make the herbicide challenged the research. But Hayes believes his science is correct. He continues to speak out about what he finds.

Wangari Muta Maathai

Wangari Muta Maathai (1940–2011) was born in Nyeri, a town in Kenya, Africa. Unlike most of the young women in her country, Maathai was able to go to college. She went to college in Kansas, Pennsylvania, and Nairobi. She was the first woman from eastern Africa to receive a doctorate degree.

In 1977, Maathai founded the Green Belt Movement. The group is mostly women. Their main activity is planting trees to replace those cut down for firewood. They have planted more than 30 million trees! But some people didn't understand the value of planting trees. They didn't see the good Maathai was doing for the environment. Even though people tried to stop her, she did what was right for the environment and for the people of Kenya.

Maathai was elected to the Kenyan parliament in 2002. She was named deputy minister of Kenyan natural resources and wildlife in 2003. In 2004, Maathai received the Nobel Peace Prize for her many years of promoting peace and good living conditions in Africa.

Range of Tolerance

Ecosystems are defined by the nonliving factors of the environment and the organisms living there. Water is a nonliving factor. Every ecosystem must have water. But the amount of water in an ecosystem can be different. Lake and ocean ecosystems thrive underwater. Rain forest ecosystems thrive with a lot of water. Desert ecosystems thrive with very little water.

Chaparral ecosystems are found on the West Coast of the United States. Chaparral is not quite as dry as desert. But plants and animals living there must survive long summers and falls without rain. The plants are tough and brushy with long roots. Many of the animals burrow deep into the rocky soil.

A chaparral ecosystem

91

Chaparral ecosystems before and after a fire

Another challenging nonliving factor in chaparral ecosystems is fire. Wildfires leave the land's surface black and lifeless. But before long, life returns. Animals that hide deep in their burrows come back to the surface after the fire passes. The roots of chaparral plants are still alive. As soon as the rains come, new branches and leaves sprout. The ashes from the burned plants provide nutrients for the new plants to grow and thrive. The chaparral ecosystem can survive well even when wildfire burns it to the ground.

The chaparral ecosystem has plants and animals that can thrive even when there are fires. The plants and animals that live there have a high range of tolerance for heat and fire. Plants and animals whose optimum environment is a forest ecosystem might survive in the chaparral ecosystem, but not nearly as well. They do not have as much tolerance for heat and fire. Plants and animals that live in rain forests would die in the chaparral ecosystem. They have no tolerance for heat and fire.

Thinking about Range of Tolerance

All plants need water. What does *optimum water* mean for a plant? What does *range of tolerance for water* mean for a plant?

How Organisms Depend on One Another

Animals depend on plants for survival. Trees provide shelter for birds to build nests. High in the branches, eggs and baby birds are safe from snakes, skunks, and coyotes. The owl in the picture below is protected from weather and predators in a tree. Beetles and isopods live under tree bark. Walking sticks hide on trees to protect themselves from predators. Animals also depend on plants for food. Animals eat leaves, flowers, fruits, seeds, bark, stems, sap, and roots of plants. It's easy to find many ways that animals depend on plants for survival.

Plants depend on animals for survival, too. You read about the swollen-thorn acacia tree. The ants help the tree survive. If an insect lands on the tree, the ants will attack it. If another plant touches the tree, the ants cut it away. The acacia tree depends on the ants for protection. And the ants depend on the tree for shelter and food.

An owl nesting in a tree

A walking stick looks like a twig.

A bee collects pollen and nectar for food.

Pollination

What other ways do plants depend on animals? Think about honeybees visiting flowers. Bees collect pollen and nectar from flowers. This is food for the bees. The bees depend on plants for food.

The plants also depend on the bees. Pollen must get from one flower to another for plants to make seeds. This is called **pollination**. Plants can't move, so the pollen must be carried from one flower to another. Bees carry pollen as they fly from flower to flower. (Can you see the yellow dust on this bee's body? That's the pollen.) Bees make it possible for plants to produce seeds. The seeds grow to become adult plants, which make flowers with pollen and nectar. Then the cycle starts over again.

Other insects, such as butterflies and moths, also visit flowers for food. Plants depend on insects to bring pollen, and insects depend on plants for food. Without bees and other insects that visit flowers, plants cannot survive. Without flowers on plants, bees and butterflies cannot survive.

Seed Dispersal

When seeds are ripe, they are ready to grow. Seeds have a better chance of survival if they sprout away from the parent plant. The new plant will be able to get more light, water, and nutrients. **Seed dispersal** is the term used to describe ways that seeds move away from the parent plant.

Sometimes wind disperses, or scatters, seeds. Wind is dispersing the small seeds of this dandelion.

Dandelion seeds blowing in the wind

Animals can also disperse seeds. Squirrels, chipmunks, and birds often take seeds and fruits (acorns, sunflower seeds, berries, and cherries) for food. They may drop the seeds or bury them and forget where they put them. Seeds with hooks can also stick to animals to be carried away from the parent plant.

Animals depend on plants for survival. Plants give animals food and shelter. Plants also depend on animals for survival. Animals help pollinate plants and disperse seeds.

Sometimes birds drop the seeds they are carrying.

A chipmunk eating an acorn

Thinking about Dependence

1. Describe three examples of how animals depend on plants for survival.

2. Describe three examples of how plants depend on animals for survival.

3. Do you think animals pollinate flowers and disperse seeds on purpose or by accident? Explain why you think so.

Animals from the Past

Thirty thousand years ago, people did not live in Los Angeles, California. But animals did. Some of these animals were the same ones you might see in the western United States today. Coyotes, mountain lions, and black bears lived back then. But some of the animals that were alive then are gone today. How do we know those animals lived 30,000 years ago?

Not far from downtown Los Angeles is a place called the La Brea Tar Pits. The area has many pools of hard, black tar. In the summer heat, the tar melts.

When animals stepped into the melted tar thousands of years ago, they got stuck. They couldn't get out. Slowly, they sank into the sticky tar and died.

In 1901, scientists discovered that the tar pits were full of **fossil** bones. Some of the bones were unlike any they had seen before. They dug the bones out of the tar. The scientists carefully put them together to make complete skeletons. Some skeletons were from kinds of animals that no longer live on Earth. Groups of animals that once lived on Earth, but have died out are extinct.

A tar pit at La Brea

La Brea Tar Pits Museum

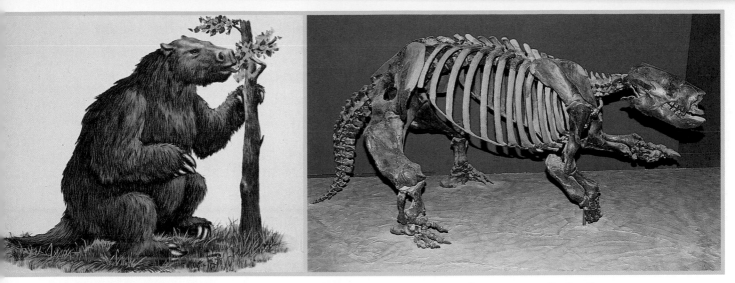

What a ground sloth might have looked like

A fossil skeleton of a ground sloth

Ground sloths were common 30,000 years ago in the western United States. Ground sloths are extinct now. Scientists discovered that ground sloths once lived in this region when they found their bones in the tar pits. The ground sloths are similar to tree sloths that still live in the rain forests of South America.

Saber-toothed cats no longer live on Earth. They are similar to the mountain lions that live in the western United States and other places today. Saber-toothed cats had huge canine teeth that look deadly. Scientists aren't sure what purpose the oversized teeth served. The large cats may have followed prey animals into the tar.

What a saber-toothed cat might have looked like

A skull of a saber-toothed cat

The mastodon is one of the largest animals that lived in this region. It looks a lot like the elephants that live in Africa and Asia. Some of the tar pits might have had a layer of water over the tar. Mastodons that stepped into the pool for a drink could have become trapped in the tar.

Why did the ground sloth, saber-toothed cat, and mastodon become extinct? It's not because they all fell into the tar pits. Only a few were trapped in the tar and died. Organisms become extinct when the environment changes.

The climate was warming up after the ice age. When the environment changed, some animals survived, some animals looked for other places to live, and some died. The animals that are known only as fossils from the La Brea Tar Pits could not survive in the changed environment.

What a mastodon might have looked like

A fossil skeleton of a mastodon

Recent Discoveries about Ice Age Animals

Ice age fossils have been discovered recently in other parts of North America. In 1978, the fossil remains of Columbian mammoths were discovered near the Bosque River in Waco, Texas. This is the same kind of mammoth that was found in the La Brea Tar Pits. Researchers from Baylor University worked for 20 years to uncover the fossils and preserve them. The researchers found 22 mammoths, a camel, and the tooth of a young saber-toothed cat. The Waco Mammoth Site holds the record for the most skeletons of mammoths that died at the same time. At least 19 of the mammoths were trapped by flood water some 68,000 years ago. Their remains were buried in the sediments.

The Waco Mammoth Site was closed to the public until the end of 2009. That year, a shelter was completed to protect the bones. Now the fossils can be viewed by the public. There are plans to make this site a national monument, like the Dinosaur National Monument in Colorado and Utah.

A Columbian mammoth

More ice age fossils turned up in October 2010 near Aspen, Colorado. A construction worker was using a bulldozer to expand the Ziegler Reservoir. He was digging into the ground and removing peat moss when he observed a few large rib bones poking out. He immediately stopped and called the Denver Museum of Nature and Science.

A team of paleontologists went to the site. They used hand tools and began to carefully uncover the well-preserved bones of a Columbian mammoth. The bones had been buried in a lake during the ice age. The lake was gone, but the sediments were still wet and covered with peat moss. The scientists quickly wrapped the bones in plaster so that the bones would stay moist. If the bones dried out too fast, they would break.

This site is one of the richest fossil sites ever discovered in Colorado. The fossils found so far include American mastodons, Columbian mammoths, ice age deer and bison, a ground sloth, and a tiger salamander. Small animals include iridescent beetles and other insects, snails, and microscopic crustaceans. The large quantity of well-preserved plant material includes wood, seeds, cones, leaves of white spruce, sedges, and other plants.

The scientists could only work for a month at the reservoir fossil site because of the winter snow. Scientists continue to return to this site during good weather, uncover more fossils, and learn more about the animals that lived there 40,000 years ago.

Uncovering the bones of a mammoth

Wrapping the fossils in plaster

This discovery is important because the elevation is almost 2,700 meters (m). Scientists have never found ice age fossils at such a high elevation before. Little is known about the animals living in high-elevation environments during the ice age.

The other important discovery is nearly two dozen mammoth and mastodon tusks. The tusks have growth rings just like tree trunks. Scientists can use the tusk growth rings to tell how old the animal was when it died. They can even tell what season of the year the animal died. The tusks become a record of the animals' lives. They tell if the female had calves, if the males had been fighting, and other interesting things about their lives. These tusks and other fossils are on display at the Denver Museum of Nature and Science.

Scientists are learning a lot about ice age environments from the fossils found at different sites. They are a way to study animals that lived in a colder environment than exists today. The information might help us understand what will happen as Earth gets warmer.

Removing the bones of a Columbian mammoth from the ground

A close-up of the mastodon tusk in the ground

Thinking about the Past

1. What does *extinct* mean?
2. What are some animals that once lived in the United States, but are now extinct?
3. What are some animals that are similar to animals that are now extinct?
4. What can cause animals to become extinct?

The tip of the tusk

Darkling Beetles

Darkling beetles are insects. They live in almost every part of the world, from the desert to the rain forest. There are many different kinds. In North America alone, there are 1,400 kinds of darkling beetles! One kind of darkling beetle is *Tenebrio*.

The adult *Tenebrio* beetle is about 1.9 centimeters (cm) long. It is dark brown to black and usually lives in dark, dry places. Like other insects, the darkling beetle has six legs and three body parts. These parts are the head, thorax, and abdomen. Like other beetles, it has two pairs of wings. The front wings cover and protect the back wings and abdomen. Even with wings, darkling beetles cannot fly.

Life Cycle

The darkling beetle goes through four stages in its life cycle. The stages are egg, larva, pupa, and adult beetle. Female beetles lay 500 to 1,000 eggs at a time. The eggs at 1 millimeter (mm) are almost too small to see. Tiny larvae hatch from the eggs in about a week.

The larvae of *Tenebrio* beetles are a yellow-gold color. They are called mealworms, but they are not worms at all. The larvae eat cereals and grains. They grow to a length of 3 cm. The larvae molt (shed their tough outer skin) several times in order to grow. After about 3 months, the larvae change into pupae.

The pupa is a resting stage. The insect's body begins to change into an adult beetle. The pupa stage lasts about 2 weeks. Then the beetle comes out as an adult. This cycle of changes is called complete metamorphosis.

A larva (mealworm)

A pupa

An adult *Tenebrio* beetle

Characteristics

Darkling beetles inherit most of their characteristics from their parents. Darkling beetles get their size and color from their parents. They get their head, antennae, thorax, and six legs from their parents.

Some characteristics are caused by the environment. Things can happen to change how a beetle looks. If a beetle gets into a fight, it might lose a piece of wing cover. It could even lose a leg. The beetle looks different.

If the beetle becomes a parent, what will its offspring look like? Will they have broken wing covers and five legs? No. Changes like these are caused by the environment. They are not passed on to offspring.

In the natural environment, *Tenebrio* beetles live in grasslands where there are plenty of seeds. They also make their homes near humans. They get into cupboards, pantries, and chicken farms. For this reason, darkling beetles might be thought of as pests. But they are harmless to humans.

Darkling beetles inherit most of their characteristics from their parents.

Other Beetles

What makes a beetle a beetle? The most important characteristic that all beetles share is their short, hard front wings called elytra. When a beetle folds its wings, the elytra cover its entire abdomen. This shell gives a beetle its armored appearance. When a beetle flies, it lifts its elytra so that its back wings can move.

All beetles go through the same four stages of growth as the darkling beetle. Females lay eggs that hatch into wormlike larvae. The larvae eat, grow, and pupate. Finally, the pupae change into adults. At least 250,000 kinds of beetles have been described by scientists. Beetles can be less than 1 cm to more than 15 cm long.

Another kind of darkling beetle

Beetles live in just about every environment on Earth. They live in rain forests, deserts, mountain lakes, rivers, and northern forests. They can live in people's homes and gardens. They can even live in sewers. The only environment they don't live in is the ocean.

Beetles eat almost everything. Some eat leaves, fruit, bark, seeds, and grains. Others are parasites and live on or in living animals. Some beetles are scavengers, living on dead animals or dung. Beetles can be helpful to humans. For example, beetles called ladybugs are predators. They eat small insects that destroy gardens and farm plants.

Beetles have antennae for sensing their environment. These antennae are primarily used to smell, but may also be used to feel.

A metallic-green fig beetle

A ten-lined June beetle

What group of insects do you think is the most successful on Earth? Flies? Mosquitoes? Ants? It's the beetles. There are more kinds of beetles than all the other kinds of insects added together.

And how many different kinds of insects are there? No one knows for sure. About 1 million of the 1.3 million kinds of organisms that have been described by scientists are insects. The list of insects is growing at the rate of about 7,000 to 10,000 new kinds every year! Based on work done in rain forests, some scientists think there may be 10 to 30 million more kinds of insects to discover. The estimated numbers of kinds of plants, fish, birds, reptiles, and mammals seem quite small compared to the millions of insects.

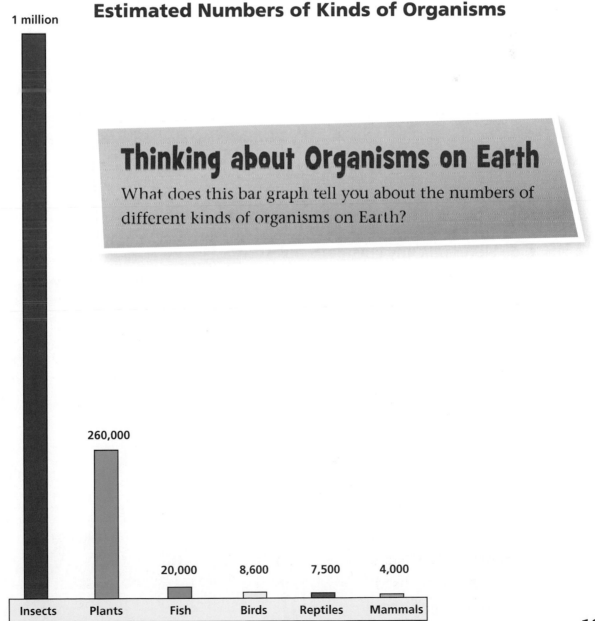

Estimated Numbers of Kinds of Organisms

1 million

260,000

20,000

8,600

7,500

4,000

Insects Plants Fish Birds Reptiles Mammals

Thinking about Organisms on Earth
What does this bar graph tell you about the numbers of different kinds of organisms on Earth?

Science Safety Rules

1. Listen carefully to your teacher's instructions. Follow all directions. Ask questions if you don't know what to do.

2. Tell your teacher if you have any allergies.

3. Never put any materials in your mouth. Do not taste anything unless your teacher tells you to do so.

4. Never smell any unknown material. If your teacher tells you to smell something, wave your hand over the material to bring the smell toward your nose.

5. Do not touch your face, mouth, ears, eyes, or nose while working with chemicals, plants, or animals.

6. Always protect your eyes. Wear safety goggles when necessary. Tell your teacher if you wear contact lenses.

7. Always wash your hands with soap and warm water after handling chemicals, plants, or animals.

8. Never mix any chemicals unless your teacher tells you to do so.

9. Report all spills, accidents, and injuries to your teacher.

10. Treat animals with respect, caution, and consideration.

11. Clean up your work space after each investigation.

12. Act responsibly during all science activities.

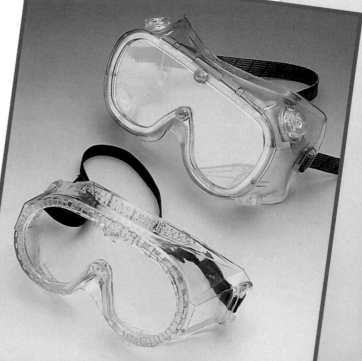

Glossary

algae a large plantlike group of water organisms

amphibian an animal, such as a frog or spadefoot toad, that reproduces in water

aquatic referring to water

bacteria microorganisms that act as decomposers

behavior the actions of an animal in response to its environment

burrow a hole or tunnel dug by a small animal

canopy the highest layer in a forest, where there is a lot of sunlight

carnivore an animal that eats only animals

climate the average or typical weather conditions in a region of the world

community the plants and animals in an ecosystem

compete to rely on or need the same resource as another organism

complete metamorphosis the cycle of growth changes for an insect. The stages include egg, larva, pupa, and adult.

concentration the amount of a substance, such as salt, in an amount of another substance, such as water

consumer an organism that cannot make its own food. Consumers eat other organisms.

crustacean a class of mostly aquatic animals with hard, flexible shells

decomposer an organism that breaks down plant and animal material into simple chemicals

dormant inactive or resting

ecologist a scientist who studies ecosystems

ecosystem a community of organisms interacting with each other and with the nonliving environment

endangered at risk of becoming extinct

energy what allows organisms to grow and move

entomologist a biologist who studies insects

environment everything that surrounds and influences an organism. Deserts, forests, and the ocean are environments.

environmental factor one part of the environment. An environmental factor can be nonliving, such as water, light, and temperature. It can be living, such as plants and animals.

evaporate to dry up and go into the air

extinction a species that no longer exists

fertile able to support growth and development

fertilizer any natural or synthetic material used in soil to help plants grow

food a form of chemical energy that organisms need to survive

food chain a description of the feeding relationships between organisms in an environment

food web all of the connected and interacting food chains in an ecosystem. Arrows show the flow of matter and energy from one organism to another.

fossil any remains, trace, or imprint of animal or plant life preserved in Earth's crust

function an action that helps a plant or an animal survive

fungus (plural **fungi**) an organism that lacks chlorophyll and gets nutrients from dead or living organisms

generation a group of organisms born and living at the same time

habitat the natural environment of a plant or an animal

herbicide a chemical used to kill plants

herbivore an animal that eats only plants or algae

hypersound a very high frequency sound that is too high for human ears to detect

inherited trait a characteristic that is passed down from generation to generation

insect an animal that has six legs, a head, a thorax, and an abdomen

interact to act upon one another

isopod a small crustacean with 14 legs that all function the same

larva (plural **larvae**) the wormlike early stage in the life cycle of an insect

living the condition of being alive

matter anything that has mass and takes up space

microorganism a microscopic organism, such as bacteria and some algae

migrate when animals move from place to place with a change in the weather

nocturnal active at night

nonliving referring to something that has never been alive or to things that were once alive and are no longer alive

nutrient a material needed by a living organism to help it grow and develop

omnivore an animal that eats both plants and animals

optimum most favorable to growth, development, and reproduction of an organism

organism any living thing

parasite an organism that lives on and gets nutrients from another living organism

pesticide a chemical developed to kill animals that are in some way harmful to humans

petroleum an oil that comes from the earth

photosynthesis a process used by plants and algae to make sugar (food) out of light, carbon dioxide, and water

phytoplankton microscopic plantlike organisms in aquatic environments that produce their own food

pollination the moving of pollen to the female part of a flower

pollute to make an environment unsuitable for organisms because of substances introduced into air, water, or soil

predator an animal that hunts and catches other animals for food

prey an animal eaten by another animal

producer an organism, such as a plant or algae, that makes its own food

pupa (plural **pupae**) the stage of an animal's life cycle between the larva and the adult stages

range an amount of variation or difference

range of tolerance the varying conditions of one environmental factor in which an organism can survive

recycle to use again

reproduce to have offspring

scavenger an animal that eats dead organisms

seed dispersal the movement of seeds away from the parent plant

senses information received from the environment using hearing, touch or feel, sight, smell, and taste

sensory receptor a specialized cell that gets information from the environment and sends it to the brain

source the beginning of something, such as where a river starts

species a group of organisms that are all the same kind

stridulation the chirping sound that crickets make

structure any identifiable part of an organism

temperature a measure of how hot or cold matter is

terrarium a container with plants growing inside

terrestrial referring to land

thrive to grow fast and stay healthy

ultrasound a very low frequency sound that is too low for human ears to detect

understory the layer above the rain forest floor and below the rain forest canopy

variation difference

vernal pool a shallow, temporary pond

zooplankton microscopic animals in aquatic environments

Index